The Aqueous Chemistry of Polonium and the Practical Application of Its Thermochemistry

The Aqueous Chemistry of Polonium and the Practical Application of Its Thermochemistry

SUSAN A. BROWN
Australian Nuclear Science and Technology Organisation, Sydney, NSW, Australia

PAUL L. BROWN
Rio Tinto Growth and Innovation, Melbourne, VIC, Australia

ELSEVIER

Elsevier
Radarweg 29, PO Box 211, 1000 AE Amsterdam, Netherlands
The Boulevard, Langford Lane, Kidlington, Oxford OX5 1GB, United Kingdom
50 Hampshire Street, 5th Floor, Cambridge, MA 02139, United States

British Library Cataloguing-in-Publication Data
A catalogue record for this book is available from the British Library

Library of Congress Cataloging-in-Publication Data
A catalog record for this book is available from the Library of Congress

ISBN: 978-0-12-819308-2

For Information on all Elsevier publications
visit our website at https://www.elsevier.com/books-and-journals

Publisher: Susan Dennis
Acquisition Editor: Anneka Hess
Editorial Project Manager: Andrea Dulberger
Production Project Manager: Surya Narayanan Jayachadran
Cover Designer: Matthew Limbert

Typeset by MPS Limited, Chennai, India

Contents

1. Polonium **1**

References 5

2. Physical and chemical properties **7**

2.1 Isotopes of polonium 7
2.2 Elemental polonium 8
2.3 Oxidation states of polonium 10
2.4 Polonium oxides, hydroxides, and hydrides 15
2.5 Polonium halides 18
2.6 Polonides, polonites, and the polonium compounds with other chalcogens 25
2.7 Polonium nitrates 29
2.8 Solvent extraction of polonium 31
2.9 Polonium ion exchange 33
2.10 Other behavior 36
References 38

3. Chemical thermodynamics of polonium **43**

3.1 General principles 43
3.2 Thermochemical properties for polonium species 52
References 116

4. The pH-potential diagram for polonium **121**

4.1 Introduction 121
4.2 The polonium–water system 122
4.3 Construction of pH-potential diagrams 122
References 126

5. The use of pH–potential diagrams in practical applications **127**

5.1 Introduction 127
5.2 Derivation of pH–potential diagrams 127
5.3 The aqueous speciation of polonium, selenium, tellurium, and lead 128
5.4 Case study—polonium behavior during anode slimes processing 130
5.5 Case study—polonium behavior during silver and gold electrorefining 163

5.6 Case study—polonium in seawater 165
5.7 Case study—autodeposition of polonium 168
5.8 Case study—the mineral processing of rare earth minerals 170
References 176

6. Conclusions **179**

Reference 180

Appendix 1: Thermochemical data *181*
Appendix 2: Polonium hydroxochloride complexation *191*
Index *193*

CHAPTER 1

Polonium

Polonium was discovered in 1898 by Pierre and Marie Curie following the observation that the radioactivity of pitchblende was about five times greater than expected on the basis of its uranium content. Large quantities of pitchblende from Joachimsthal were processed and an intensely radioactive substance was carried down with bismuth sulfide precipitated from hydrochloric acid solution. Tracer experiments indicated that the radioactive substance was a new element, polonium, named after Marie Curie's birth country, Poland (Wahl and Bonner, 1951). Four years later, Marckwald demonstrated that the radioactivity that had been concentrated with bismuth could be separated from the latter using both cathodic and spontaneous deposition and by precipitation from aqueous solution using stannous chloride (Bagnall, 1957). The name suggested for the enhanced radioactive substance was radio-tellurium, due to the chemical resemblance of the separated material to tellurium. Within 2 more years, Rutherford identified, from the nature of the residual activity observed from the decay of radon, a substance that was identified as radium-F (originally identified as radium-E) (Bagnall, 1957; Fry and Thoennessen, 2013). Rutherford later identified that polonium, radio-tellurium, and radium-F were all the same, being one of the daughter products of ^{238}U, the isotope ^{210}Po of the new element polonium.

There are a large number of isotopes of polonium, but only seven are produced within the uranium, actinium, and thorium decay chains that occur naturally. The longest lived naturally occurring isotope of polonium is ^{210}Po with a half-life of 138.378 days (Kocher, 1977). Polonium-210 is very dangerous to handle as a result of intense radiation release and requires special equipment and strict control even with milligram or microgram quantities. The longest lived isotope is ^{209}Po [$t_{1/2} = 102$ years (Fry and Thoennessen, 2013)], which is synthetic and is produced by bombardment of lead or bismuth in a cyclotron.

The other six naturally occurring polonium isotopes were all discovered within 20 years of the discovery of ^{210}Po. They were all identified from their unique radioactivity decay energies. Synthetic polonium isotopes could not be produced until the development of cyclotrons and

The Aqueous Chemistry of Polonium and the Practical Application of its Thermochemistry
DOI: https://doi.org/10.1016/B978-0-12-819308-2.00001-2

nuclear reactors, almost 50 years after the identification of polonium. The most recent isotopes, ^{223}Po to ^{227}Po, were first produced as recently as 2010 (Fry and Thoennessen, 2013).

Polonium is about eight orders of magnitude more toxic than hydrogen cyanide (McFee and Leiken, 2009). However, polonium is only a health hazard if it is taken internally. External exposure does not represent an issue since polonium is an alpha emitter (although it does have photon emissions, these are extremely low in abundance) (Ansoborlo et al., 2012). Therefore polonium is only a health issue if it is inhaled or ingested. Polonium-210 emits alpha particles of high energy (5 MeV) which are capable of traveling around 50 μm in water and biological tissues. Cells in the human body are typically between 10 and 30 μm in diameter and, as such, alpha particles released from the decay of polonium will have devastating effects on cell structures and DNA (Ansoborlo et al., 2012). Polonium-210 toxicity causes symptoms which are similar, but not identical, to acute radiation syndrome that is caused by whole body gamma-radiation. These symptoms are bone marrow syndrome, gastrointestinal syndrome, and central nervous system syndrome (Ansoborlo et al., 2012). After ingestion or inhalation, polonium is deposited primarily in soft tissues, with the greatest concentrations in the reticuloendothelial system, principally the liver, spleen, and bone marrow, as well as in the kidneys and skin (hair follicles) (McFee and Leiken, 2009). Typically, four phases of illness are followed: prodrome including nausea and vomiting (which occurs in minutes to days), hematopoietic latency (none to weeks) where white cells and platelets decrease, illness (days to weeks), and finally death or recovery (weeks to months). The severity of these symptoms will depend on the extent of exposure and the ability to detect the exposure and administer an appropriate chelator drug quickly enough (McFee and Leiken, 2009). The effects of polonium are similar in a range of animals including man. The lethal dose (LD_{50}) has been determined to be $6-15\ ng\ kg^{-1}$ body mass. Thus for an 80 kg human, the lethal dose is approximately 1 μg (this amount of polonium would initially emit in excess of 25 billion alpha particles every second).

There have been a number of cases of polonium poisoning. The first documented case was that of Cotelle, who was a technician of the Curie's, and was working with Irene Joliot-Curie when a vessel containing polonium exploded. She inhaled a lethal dose of polonium and died 2 weeks later (Ansoborlo et al., 2012). Joliot-Curie was standing behind Cotelle and felt no immediate effects; however, she eventually contracted leukemia and died 10 years later as a consequence of the polonium exposure. In another incident, a Russian worker inhaled an estimated dose of 530 MBq. Symptoms included vomiting, severe fever, and a decrease of red blood cells. The worker died 13 days after contamination (Ansoborlo et al., 2012). The most famous case occurred in 2006. Alexander Litvinenko, a former Russian spy, was poisoned with polonium in a London restaurant. It was not initially recognized that polonium had been administered, and consequently, it took some period of time before the correct tests were carried out that revealed substantial contents of alpha radiation in a urine sample in the form of ^{210}Po. Litvinenko died a few days later. It has also been suspected that the Palestinian leader Yasser Arafat may have been poisoned with polonium, but no conclusive evidence for this exists.

Polonium-210 also occurs in tobacco. The presence of polonium results from the widespread usage of phosphate fertilizers which naturally contain some uranium. Tobacco grown in countries such as Turkey, India, and Indonesia, where organic fertilizers are used, contain a reduced concentration of polonium (Rego, 2009). The polonium contained in tobacco is volatized into smoke, and consequently, can enter the smoker's lungs. Research conducted in the 1960s demonstrated that polonium in smoke concentrated in branching points of the bronchial epithelium (so-called hot spots) that only account for 2%–3% of the lung (Rego, 2009). Tests conducted on hamsters with polonium forced into the trachea showed that 94% developed lung tumors with such small doses that inflammation did not occur. A later study on the same animals demonstrated that exposure to low doses of polonium resulted in 10%–36% developing malignant tumors. This compares to approximately 15% of lifelong smokers who develop lung cancer (Rego, 2009). Consequently, it appears certain that the presence of polonium in tobacco is one significant cause of the development of cancers in smokers.

Polonium has few applications; however, as a consequence of its intense radiation, polonium has been used as a thermoelectric generator and in satellites and moon rovers as a heat source. Polonium is also used to eliminate static charges in photographic plates, textile mills, paper rolls, and sheet plastics; however, the polonium needs to be replaced frequently due to the short half-life of ^{210}Po.

Uranium-rich ores typically contain about 100 μg of ^{210}Po per tonne of ore. Many other ores can contain uranium and, as such, will also contain ^{210}Po, but at concentrations substantially lower than those contained in typical uranium ores. Nevertheless, when ores containing uranium are processed, polonium partitions between the solid and liquid phases according to its chemistry. This can often lead to unacceptably high levels of contamination which affect the quality of the final products and also pose an occupational health and safety risk to plant operators and maintenance staff.

Although techniques have been developed for the removal and/or control of polonium in many ore processing streams, little is known about the chemistry of the element and, in particular, its aqueous chemistry is poorly understood. The efficacy of the developed techniques could, therefore, be significantly improved by an increased understanding of the aqueous chemistry of polonium.

To understand the aqueous chemistry of any element requires the formulation of a list of species which are likely to form for a particular set of elements/ions under specified conditions. The relative formation of each individual species is dependent on its stability, the concentrations of reacting elements/ions, and solution characteristics (e.g., pH, redox potential). For polonium, until now, no such database has been available.

Most commonly, the acquisition of a thermochemical database can be achieved by assessing such data from literature sources. On occasions, however, and particularly for radioactive elements, the extent of the literature data is limited or nonexistent and often inconsistent. This can be overcome, to some degree, by the use of thermochemical data derived from the use of theoretical equations aimed solely at predicting such data. These equations are validated, in the first instance, by comparison of predicted and literature data, after which the equations can be used to predict thermochemical data of other species.

Once the species list has been established, the data, together with its concomitant suite of thermochemical parameters, can be used to predict the speciation characteristics of the element for a given set of solution conditions. These characteristics can be used before a particular experiment is conducted to indicate likely outcomes or to assess, for example, a particular industrial process. In industry, such information is often essential in understanding the environmental consequences of a process.

Uranium can be present in a number of types of ores. The processing of these ores can often concentrate polonium leading to potential issues with elevated concentrations in products and process streams (e.g., fumes, due to the high volatility of polonium). For example, there are a number of significant global copper producers where the presence of uranium can be problematic. These operations often have integrated processes to produce high-quality refined copper and, at times, refined gold and silver and, more rarely, uranium oxide as coproducts. An example of such an operation is BHP's Olympic Dam Operations at Roxby Downs, South Australia, which is the eighth largest copper and the largest uranium ore body in the world.

Polonium concentrations in process liquors at some copper operations have, from time to time, resulted in unacceptably high contamination levels in the final products. Companies running these operations have, in the past, undertaken projects to address specific problems in several areas of their plant circuits. These investigations were in general successful, and they served to highlight the lack of understanding of the chemistry of polonium in the process circuits. Although polonium is strictly controlled, operating experience has shown that polonium concentrations can vary significantly, and the reasons for these changes are poorly understood. It follows that in areas of the plant circuit where concentrations are close to allowed limits, changes to process conditions could result in unacceptable polonium levels.

Rare earth mineral concentrates, particularly those of monazite, can also contain thorium and uranium and, as such, will also contain polonium. There is limited understanding of the deportment of radionuclides in the processing of minerals containing rare earth elements. As with copper operations, where uranium is a by-product or is present in quantities that need careful management, companies processing rare earth element concentrates will also need to implement strict control practices so that the final products contain radioactive concentrations lower than regulatory limits.

To be able to predict the behavior of polonium and to develop effective methods for its removal or control, it is necessary to acquire an understanding, or indication, of its aqueous speciation in these types of circuits. A study has been undertaken to address these issues. A thermochemical database for polonium has been developed, and then used to derive pH-potential diagrams for various processing circuits. The aim was to gain a better understanding of polonium behavior in these circuits which would ultimately allow them to drive polonium in the most appropriate direction, as dictated by process needs.

References

Ansoborlo, E., Berard, P., Den Auwer, C., Leggett, R., Menetrier, F., Younes, A., et al., 2012. Review of chemical and radiotoxicological properties of polonium for internal contamination purposes. Chem. Res. Toxicol. 25, 1551–1564.

Bagnall, K.W., 1957. Chemistry of the Rare Radioelements. Butterworths Scientific. Publications, London.

Fry, C., Thoennessen, M., 2013. Discovery of actinium, thorium, protactinium, and uranium isotopes. Atom. Data Nucl. Data Tables 99, 345–364.
Kocher, D.C., 1977. Nuclear Decay Data for Radionuclides Occurring in Routine Releases From Nuclear Fuel Cycle Facilities. Oak Ridge National Laboratory, ORNL/NUREG/T102.
McFee, R.B., Leiken, J.B., 2009. Death by polonium-210: lessons learned from the murder of former Soviet spy Alexander Litvinenko. Semin. Diagn. Pathol. 26, 61–67.
Rego, B., 2009. The polonium brief. A hidden history of cancer, radiation and the tobacco industry. Isis 100, 453–484.
Wahl, A.C., Bonner, N.A. (Eds.), 1951. Radioactivity Applied to Chemistry. John Wiley & Sons, New York.

CHAPTER 2

Physical and chemical properties

2.1 Isotopes of polonium

There are only seven naturally occurring isotopes of polonium; two in the thorium ($4n$) decay series (^{212}Po and ^{216}Po), two in the actinium ($4n+1$) decay series (^{211}Po and ^{215}Po) and three in the uranium ($4n+2$) decay series (^{210}Po, ^{214}Po, and ^{218}Po). The two isotopes in the thorium chain have very short half-lives, with the longest being that for ^{216}Po of 0.145 seconds (Fry and Thoennessen, 2013). Polonium-211 and ^{215}Po are also both short-lived, with both again having half-lives of less than 1 second. In comparison, two of the uranium series isotopes are much longer lived (^{210}Po and ^{218}Po), with half-lives of 138.378 days (Kocher, 1977) and 3.098 minutes (Fry and Thoennessen, 2013), respectively. The half-life of the third polonium isotope in the uranium series (^{214}Po) is very short. Polonium-210 was the form of polonium discovered by the Curies in 1898. Four years later radio-tellurium was identified, but this was later shown to be identical to ^{210}Po (Fry and Thoennessen, 2013).

In addition to the seven natural isotopes, there are a further 35 synthetic isotopes of polonium that have been produced. These are ^{186}Po to ^{209}Po, ^{213}Po, ^{217}Po, and ^{219}Po to ^{227}Po (Fry and Thoennessen, 2013). The vast majority of these isotopes have very short half-lives, but two, ^{208}Po and ^{209}Po, have half-lives longer than any of the naturally occurring polonium isotopes. The currently accepted half-lives of these two isotopes are 2.898 and 102 years, respectively (Fry and Thoennessen, 2013). As a consequence of the relatively long half-life of ^{209}Po, it is used as a tracer in the analysis of natural polonium. Polonium-208 was initially produced by the bombardment of ^{207}Pb with alpha particles (helium) and ^{209}Po by the bombardment of bismuth with deuterons, with both isotopes produced in a cyclotron.

Due to the short half-lives of naturally occurring polonium, the amount that could be separated in initial experimental work was exceedingly small and its presence could only be followed by its radioactivity (Bagnall, 1957b). Large amounts of uranium ore residues were processed by the Curies and a product was produced that was orders of magnitude

The Aqueous Chemistry of Polonium and the Practical Application of its Thermochemistry
DOI: https://doi.org/10.1016/B978-0-12-819308-2.00002-4

more radioactive than uranium. This product, containing ^{210}Po, could be concentrated in a bismuth fraction and precipitated from solution using hydrogen sulfide (Bagnall, 1957b). Polonium sulfide was found to be more volatile than bismuth sulfide and the two sulfides could be separated using vacuum submilation. Thus polonium (as ^{210}Po) had been discovered.

2.2 Elemental polonium

2.2.1 Physical properties

Elemental polonium is reported to be metallic and resembles lead and bismuth in its physical properties, while chemically it is similar to the sulfur group elements, selenium and tellurium (Bagnall, 1957a). The electron configuration of neutral polonium atoms in their ground state is probably $(Xe)\cdot 4f^{14}5d^{10}6s^{2}6p^{4}$ $(^{3}P_{2})$, analogous to selenium and tellurium. The melting point of α-polonium is 254°C (Maxwell, 1949) with a density of 9.196 g cm^{-3} (Goode, 1956). The melting point and boiling point (962°C) are much lower than the corresponding values for tellurium and are comparable to those of thallium, lead, and bismuth (Bagnall, 1983). The boiling point has been extrapolated from the vapor pressure data of Brooks (1955). Brooks also determined that the vaporization enthalpy has a value of 102.9 ± 0.1 kJ mol^{-1}. Some properties of polonium are listed in Table 2.1.

X-ray diffraction (XRD) studies have indicated that the metal exists in at least two crystalline forms; "low-temperature" α-polonium with a simple cubic lattice and "high-temperature" β-polonium with a simple rhombohedral lattice. The lattice parameter of α-polonium was measured by Beamer and Maxwell (1949) and was found to be $a = (3.345 \pm 0.002)$ Å. The same authors also measured the lattice parameter of β-polonium, with values determined of $a = (3.359 \pm 0.002)$ Å and an angle $\alpha = 98° 13' \pm 3'$. The phase transformation occurs at about 75°C (Beamer and Maxwell, 1949). Stull and Sinke (1956) indicated that the sluggish nature of the transition suggests a small heat of transition and, as such, it can be neglected. The density of β-polonium is 9.398 g cm^{-3} (Goode, 1956). The densities of both α-polonium and β-polonium were indicated to have uncertainties of 0.006 g cm^{-3} and were based on X-ray data, the unit cell volume, and the number of atoms in the unit cell (Goode, 1956). Both crystal modifications of the element are metallic in character with a positive temperature coefficient of resistivity in contrast to sulfur, selenium, and tellurium.

Table 2.1 Some properties of polonium.

Property	Value	Reference
Atomic number	84	
Standard state	Solid at 298K	
Color	Silvery	
Crystal structure	Cubic at 298K	
Electronic configuration	$(Xe)\cdot 4f^{14}5d^{10}6s^2\cdot 6p^4$ (ground state)	
Melting point	527K	Maxwell (1949)
Boiling point	1235K	Bagnall (1983)
Solid density	9.196 g cm^{-3} (α-polonium)	Goode (1956)
Molar volume	22.54 cm^3 (α-polonium)	
Electron affinity	183.3 kJ mol^{-1}	Winter (2000)
Ionization enthalpy (first)	813 kJ mol^{-1}	Greenwood and Earnshaw (1998)
Bond length (Po–Po)	334.5 pm (α-polonium)	Beamer and Maxwell (1949)
Atomic radius	164 pm	Beamer and Maxwell (1949)
Ionic radii	94 pm Po(IV) 6-coordinate	Shannon (1976)
	108 pm Po(IV) 8-coordinate	Shannon (1976)
	67 pm Po(VI) 6-coordinate	Shannon (1976)
	228.4 pm Po(II)	Bagnall (1983)
Electrical resistivity	42 μΩ cm (α-polonium)	Maxwell (1949)
Vaporization enthalpy	102.9 kJ mol^{-1}	Brooks (1955)
Atomization enthalpy	144 kJ mol^{-1}	Stull and Sinke (1956)

Polonium plated onto platinum foil from nitric acid solution has been studied using a Baird grating spectrograph. From the results, an ionization potential of 8.43 V was reported for the neutral polonium atom (Charles et al., 1955). Finkelnburg and Stern (1950) studied the change of the screening constant for the outermost electron of an atom, $Z - Z_{eff}$, from element to element, from which they determined an ionization potential for polonium of 8.4 ± 0.3 V in excellent agreement with the value of Charles et al. (1955). The electrical resistivity of α-polonium was measured by Maxwell (1949) and found to be 42 ± 10 μΩ cm.

Ionic radii for tetravalent and hexavalent polonium have been reported by Shannon (1976) being 0.94 and 0.67 Å, respectively (both six coordinates). Eight coordinate polonium(IV) has an ionic radius of 1.08 Å

(Shannon, 1976). Bagnall (1983) lists an ionic radius for the polonide ion (Po^{2-}) of 2.284 Å. The atomic radius for polonium was reported by Beamer and Maxwell (1949) to have a value of 1.64 Å.

2.2.2 Chemical properties

The metal is precipitated when polonium compounds are treated with anhydrous liquid or concentrated aqueous ammonia, or with primary or secondary aliphatic amines, a reaction that apparently results from the formation of atomic hydrogen from the ammonia or amine under alpha bombardment. Prepared in this way, the metal is obtained as a gray-black powder, whereas the vacuum-sublimed metal has a bright, silvery appearance. A common method of preparation is by the thermal decomposition of polonium sulfide (PoS) or dioxide (PoO_2) in a vacuum.

The chemistry of polonium at low concentrations (10^8–10^5 and 5000–40 atoms) was studied by Reischmann et al. (1984, 1986) using electrodeposition, coprecipitation, and chromatography techniques. Thcy obtained complete recovery of polonium over the whole concentration range in each of the systems and concluded that polonium in extremely small amounts displays normal chemical behavior. Borg and Dienes (1981) conducted a theoretical study on the validity of single atom chemistry and found that about 10 atoms should be sufficient to establish chemical identity under normal conditions.

2.3 Oxidation states of polonium

2.3.1 Chemical properties

Valencies of −2, +2, and +4 are comparatively well known and have been well established by characterization of polonides and a hydride (−2), the halides (+2), and the dioxide (+4). There is also some evidence for a +6 state and a stable, volatile hexafluoride is the most likely compound of this valency; however, it has not been prepared. There is no conclusive evidence for the +3 state characteristic of many bismuth compounds (Bagnall, 1957a).

Polonium metal is rapidly dissolved by 2 mol L^{-1} hydrochloric acid giving, first, a pale pink solution believed to contain the Po^{2+} ion. This solution gradually becomes yellow and, on evaporation, yields polonium tetrachloride. Addition of hydrogen peroxide or chlorine water to the pink solution accelerates the color change, while the yellow solution containing the tetrachloride is reduced to the pink chloride by sulfur dioxide or hydrazine in the cold, and by arsenic(III) oxide on warming. The metal reacts vigorously with concentrated nitric acid giving a yellow solution [when concentrated, 10 curies of $^{210}Po\ mL^{-1}$ (i.e., 2.2 g L^{-1})], which becomes colorless on dilution. It is probable that the concentrated solution contains tetravalent polonium nitrate (Bagnall and D'Eye, 1954).

Hydroxylamine and oxalic acid have no effect and electrochemical observations on the latter may, therefore, be due to the formation of an oxalate complex. Reduced solutions oxidize back to the tetravalent state after excess reducing agent is eliminated. The change of potential with time at a platinum electrode has been measured to determine the valency state of the reduced polonium. The values were 620 mV at 4 minutes [Po^{2+} (pink)] and 470 mV at 12 minutes [Po^{4+} (yellow)]. The break in the potential–time curve (510 mV at 7.5 minutes) could correspond to the formation of an intermediate trichloride [Po^{3+} (pink)]. However, there is no other evidence for the existence of this compound (Bagnall et al., 1955a).

Electrochemical deposition experiments have shown that polonium (IV) can be readily reduced by chemical agents such as sulfurous, oxalic, and nitrous acids (Wahl and Bonner, 1951). Bagnall (1957b) reported that polonium, unlike tellurium, was not reduced to the metal by sulfur dioxide or hydrazine. In the presence of selenium; however, polonium was precipitated quantitatively from hydrochloric or hydrofluoric acids by reduction using both hydrazine or sulfur dioxide, and it seems possible that a polonium selenide may form.

Polonium can be precipitated from solution by stannous chloride and from sulfuric acid using sulfur dioxide. Polonium is reduced in both acetic acid and alkaline solution using hydrazine in the cold, and in sulfuric acid at the boiling point whereas hydroxylamine reduces polonium to a lower oxidation state in acetic acid or sulfuric acid at the boiling point. Trivalent titanium and hypophosphorous acid have been used to precipitate polonium metal from acid solutions of salts in the cold. Sodium dithionite reduces trace polonium from acid solution, probably as the metal, while reduction with dithionite in alkaline solution in an atmosphere of hydrogen was reported to yield a polonide. Trace amounts of sodium nitrite did not react with polonium in dilute nitric acid (0.1 mol L^{-1}); however, in more concentrated solutions of the nitrite [(0.2–4) mol L^{-1}], most of the polonium present is precipitated. The nature of the precipitate is unknown. Nitrites have also been used to decompose polonium tetraiodide for analysis without the precipitation of polonium (Bagnall, 1957b).

The oxidation of tetravalent polonium with 0.5%–5% potassium permanganate in 1–8 mol L^{-1} nitric acid under reflux yields a sludge of hydrated manganese dioxide containing all the polonium. The polonium in the sludge is not dissolved by 2 mol L^{-1} hydrochloric acid, aqueous ammonia, or 2 mol L^{-1} potassium hydroxide, whereas the polonium dissolves immediately in a mixture of 2 mol L^{-1} hydrochloric acid and 20% (by volume) hydrogen peroxide, yielding a solution containing tetravalent polonium. There is no reaction when polonium hydroxide is boiled with potassium persulfate in 2 mol L^{-1} potassium hydroxide or when polonium nitrate in 2 mol L^{-1} nitric acid is boiled with 0.002 mol L^{-1} chromium trioxide. The addition of excess ceric salt in 2 mol L^{-1} nitric acid to solutions of polonium nitrate in the same acid also has no effect, and polonium hydroxide suspended in 2 mol L^{-1} potassium hydroxide is not oxidized by chlorine, in contrast to the behavior of tellurium (Bagnall et al., 1958a).

The electrode potential for Po/Po^{2-} has been calculated to be −1.0 V by extrapolation of the corresponding potentials of the other Group 6B elements. The values observed for the anodic deposition potential indicate that deposition does not occur by discharge of the Po^{2-} ion (Bagnall, 1957b). A higher oxide, probably PoO_3, is considered to be the product obtained by anodic deposition of trace polonium and chemical evidence seems to support this view. After polonium is extracted from nitric, hydrochloric, or sulfuric acid solutions into methylisobutylketone (MIBK), the addition of a strong oxidant (Ce^{4+}, Cr^{6+}) displaces the equilibrium in favor of the aqueous phase. Reduction of the oxidant reestablishes the initial distribution in the case of Ce^{4+} and leads to an intermediate distribution with dichromate, probably due to incomplete reduction. An approximate value of 1.5 V is assigned to the Po^{4+}/Po^{6+} couple (Figgins, 1961). From the observed values of the electrode potential of Po/Po^{4+}, it appears that polonium lies between tellurium ($Te/Te^{4+} = 0.63$ V) and silver ($Ag/Ag^{+} = 0.7996$ V) in the electrochemical series, which is in agreement with its behavior in solution toward reducing agents (Bagnall, 1957b; Figgins, 1961).

A table of redox potentials in various solutions proposed by a number of authors has been summarized by Moyer (1956). From data published in the literature, Wahl and Bonner (1951) proposed tentative oxidation potential schemes for polonium, assuming the lower positive state to be +2, and also a scheme for solutions at 10^{-9} mol L^{-1} at which much of the experimental work was performed.

A detailed summary of the electrochemistry of polonium has been given by Bagnall (1957b). The electrode potential, $E°$, for the Po/Po^{4+} couple was given as 0.75 V. In extremely dilute solution, the normal electrode potential could only be obtained by extrapolation of the critical deposition potential, determined from a plot of the rate of deposition against the cathode potential. For solutions of trace polonium ($\sim 10^{-9}$ mol L^{-1}) in dilute nitric acid, a value of about 0.6 V (referred to the normal calomel electrode) was obtained. Polonium was, however, expected to lie between bismuth and tellurium in the electrochemical series, and it was thought that the above value was incompatible with the failure of hydrazine to precipitate polonium from solution as the metal. Further work did not confirm this view. The critical deposition potential of polonium ($\sim 10^{-9}$ mol L^{-1}) in 0.1 normal sulfuric, nitric, or acetic acids or 1 normal phosphoric acid is 0.37 V (referred to normal calomel electrode) and it is evident that the same polonium cations are present in each of these acids. In the presence of reducing agents, this value falls to 0.10 V. Some of the work at trace levels suggested that the modified Nernst equation might be invalid (i.e., $E = E° + (RT/nF) \ln a$). Later and more precise measurements do show some deviation; however, it should be emphasized that at the very low concentrations of polonium used for measurements (10^{-8}–10^{-13} mol L^{-1}), the electrode will be only partially covered by the polonium deposit and the activity of the solution will no longer be unity. There is as yet no comprehensive and quantitative theory that will predict the electrochemical behavior of an element in such extremely dilute solutions. Representative values for electrode potentials are given in Table 2.2.

Electrode potentials for the Po/Po^{4+} couple (in dilute nitric acid), the $Po/PoCl_6^{2-}$ couple, the $PoCl_4^{2-}/PoCl_6^{2-}$ couple, and the $Po/PoCl_4^{2-}$ couple, have been obtained. Values for hydrochloric acid at 6×10^{-6}–10^{-4} mol L^{-1} polonium referred to the complex ions $PoCl_6^{2-}$ (Po(IV)) and $PoCl_4^{2-}$ (Po(II)), but could, however, be equally applicable to the corresponding complex ions $PoCl_5^-$ and $PoCl_3^-$. Bagnall (1957b) was not able to ascertain which of these complex ions were present in hydrochloric acid solution. Moyer (1956) also reported that experiments to determine the oxidation states of polonium in acid solutions were not conclusive. In hydrochloric acid, two equal reduction steps were found (i.e., +4 to +2 and +2 to 0), while only one occurred in nitric acid. Other evidence indicated that the oxidation states in hydrochloric acid were +4 and +3, with these conclusions based on the time required to complete the second oxidation reaction of polonium.

Table 2.2 The electrode potentials for polonium.

Couple	Electrode potential (V)	Reference
Acid		
Po/Po^{2+}	0.6	Wahl and Bonner (1951)
	0.38[a]	Bagnall and Freeman (1956a)
	0.65	Latimer (1952)
	0.6	Charlot (1958)
	0.68	Nikol'skii et al. (1958)
	0.37[a]	Winter (2000)
Po^{2+}/PoO_2	1.0	Wahl and Bonner (1951)
	0.72[a]	Bagnall and Freeman (1956a)
	0.8	Latimer (1952)
	1.1	Winter (2000)
Po/PoO_2	0.8	Wahl and Bonner (1951)
	0.77	Bagnall et al. (1955a)
	0.76	Bagnall and Freeman (1956a)
	0.55[a]	Bagnall and Freeman (1956a)
	0.74	Latimer (1952)
	0.724	Van Muylder (1966)
	0.73	Zhdanov (1985)
	0.73	Winter (2000)
PoO_2/PoO_3	1.5	Matsuura and Haissinsky (1958)
	1.509	Zhdanov (1985)
	1.51	Winter (2000)
Po^{2+}/PoO_3	1.3	Winter (2000)
PoH_2/Po	c. − 1.0	Winter (2000)
Base		
Po/Po^{2-}	c. − 1.4	Winter (2000)
$Po/(PoO_3)^{2-}$	−0.5	Winter (2000)
Po/PoO_3	0.16	Winter (2000)
$(PoO_3)^{2-}/PoO_3$	1.48	Winter (2000)

[a]Measurement made in HCl and refers to complex ions.

The oxidation potentials for the $Po/PoCl_4^{2-}$ couple, at 25°C, determined by Eichelberger et al. (1965) in 1.0, 1.5, 2.0, 3.0, and 4.0 mol L^{-1} hydrochloric acid were 0.417, 0.387, 0.367, 0.342, and 0.297 V, respectively. The potentials for the $PoCl_4^{2-}/PoCl_6^{2-}$ couple at 25°C in 1.0 and 1.5 M hydrochloric acid were 0.717 and 0.702 V, respectively. Bagnall and Freeman (1956a) measured a potential for this couple of 0.72 V in 1 mol L^{-1} hydrochloric acid at 22°C while Power (1949a,b) measured a potential of 0.582 V in 4.7 mol L^{-1} hydrochloric acid. The potentials obtained by Eichelberger et al. (1965) for the $Po/PoCl_6^{2-}$ couple at 25°C were 0.567 and 0.545 V in 1.0 and 1.5 mol L^{-1} hydrochloric acid, respectively, whereas a value of 0.55 V in 1 mol L^{-1} hydrochloric acid at 22°C was obtained by Bagnall and Freeman (1956a).

The cathodic deposition of Po^{2+} was studied by Joliot (1930). A potential of 0.63 V in 0.25 mol L^{-1} sulfuric acid was obtained. The potential of polonium peroxide was determined at trace levels and values of 0.89 and 0.82 V (referred to normal calomel electrode) were obtained. A value of 0.55 V was calculated for the PoO_3/PoO_3^{2-} electrode. Marked changes in the critical deposition potential of trace polonium in the presence of oxalic acid have been ascribed to the reduction of polonium to a lower valency state. It seems more likely that these results are due to complex ion formation, and migration experiments have shown that nearly all the polonium is transported to the anode in the presence of oxalic acid. It has been observed that anodic deposition of polonium is repressed by oxalic acid, although the mechanism involved is uncertain. It should be noted that trace polonium is soluble in neutral oxalate solution and hydrolysis does not appear to occur, suggesting the presence of an oxalato-complex. The presence of a singly charged complex ion, $(Po^{III}(C_2O_4)_2(H_2O)_2)^-$, has been postulated to explain the results of diffusion studies. However, polonium would probably be in the tetravalent state under the conditions of the experiment. Cocrystallization studies with lanthanum, yttrium, and scandium oxalates suggest the presence of trivalent polonium, although this seems unlikely. Further work on a milligram scale established the divalent and tetravalent states only (Bagnall, 1957b).

2.4 Polonium oxides, hydroxides, and hydrides

2.4.1 Physical properties

XRD patterns of polonium dioxide (PoO_2), prepared from the reaction of polonium metal and oxygen, and those prepared by the decomposition of polonium nitrate are identical. In both cases, it was found that the oxide existed successively in different solid phases. The first, which lasted only a few days after the compound was prepared, is tetragonal. The second had a cubic fluorite (CaF_2) type structure. The radius ratio, r/x, of the compound is 0.73, which is the lower limit of stability for the cubic configuration (Bagnall, 1957b). The cubic form has a lattice parameter of $a = 5.59$ Å and a theoretical density of 9.18 g cm^{-3} (Moyer, 1956).

2.4.2 Chemical properties

Polonium dioxide becomes chocolate brown at 885°C in oxygen at 10^5 Pa and sublimes under these conditions; however, it decomposes to the elements at about 500°C under vacuum (Bagnall, 1983). The dioxide reacts at 250°C with both ammonia gas and hydrogen sulfide to give black solids, whose compositions are not known. When polonium dioxide is heated in sulfur dioxide at 250°C, a white solid remains, presumably a sulfite, since the corresponding sulfates are purple or yellow at this temperature. The dioxide does not react with liquid sulfur dioxide (Bagnall et al., 1958a).

The dioxide is appreciably more basic in character than tellurium dioxide, forming compounds such as the disulfate ($Po(SO_4)_2 \cdot H_2O$), for which no tellurium analog is known; however, it still shows some acidic character. The hydrated dioxide, a pale yellow, flocculant precipitate obtained when dilute aqueous alkali is added to an aqueous solution containing polonium(IV), is feebly acidic and a colorless product, probably a polonite(IV), forms when the dioxide is fused with potassium hydroxide or nitrite in air (Bagnall, 1983).

Polonium dioxide is readily soluble in dilute hydrochloric or hydrobromic acids giving yellow and orange-red solutions, respectively, which yield yellow polonium tetrachloride, and carmine-red tetrabromide on evaporation to dryness. A solution of aqueous, iodine free hydroiodic acid reacts with the oxide to give a very volatile black, insoluble solid, believed to be polonium tetraiodide (Bagnall and D'Eye, 1954).

The monoxide is produced in the spontaneous decomposition of polonium sulfite, or selenite, $PoXO_3$ ($X = S$, Se). The corresponding hydrated oxide or hydroxide forms as a dark brown precipitate on the addition of alkali to solutions of the dihalides in acid. It is rapidly oxidized to the +4 state (Bagnall, 1957a).

The trioxide is thought to be formed on the tracer scale by the anodic deposition of polonium from acidic media, though it has not been fully characterized. Fusion of polonium dioxide with a mixture of potassium hydroxide and chlorate yields a bluish colored solid, which is more soluble in water than the dioxide alone and may well contain polonate(VI) (Bagnall, 1983).

Tracer solution chemistry indicates that polonium hydroxide is acidic and analogous to tellurous acid. The reported solubility for this compound in water or excess alkali of 0.075 mg of Po L^{-1} (no temperature given) (Bagnall et al., 1955a) appeared to be much larger than the values suggested by tracer work, and hence Bagnall and Freeman (1957) investigated the dependence of the solubility in 0.26–1.73 mol L^{-1} potassium hydroxide.

The solubility in alkali increased slowly for 24 hours and then remained constant, and was not affected by the addition of hydrogen peroxide. The equilibrium was approached from both higher and lower concentrations of alkali, with results consistent to within $\pm 1\%$. Activity corrections were not applied and a plot of log (KOH) versus log (solubility) was linear with a slope close to 2. Eq. (2.1) describes the probable reaction.

$$PoO(OH)_2 + 2KOH \rightleftharpoons K_2PoO_3 + 2H_2O \tag{2.1}$$

The equilibrium constant, K, is $8.2 \pm 0.4 \times 10^{-5}$ at 22°C, which suggests that polonium hydroxide is, therefore, much less acidic than tellurous acid. Like the latter, its solubility in aqueous ammonia is little different from its solubility in water (4.2×10^{-7} mol L^{-1}).

Values of 0.044 and 0.22 mg L^{-1} have been reported for the solubility of the hydroxide in 18 mol L^{-1} ammonium hydroxide and that of 3.7×10^{-5} mg L^{-1} for the solubility in 0.1 mol L^{-1} ammonium hydroxide. Values for the solubility product of $Po(OH)_4$ range between 10^{-37} and $10^{-38.2}$ (Figgins, 1961; Sillén and Martell, 1964).

From analogies with selenium and tellurium, electrical migration experiments and the increasing solubility of polonium hydroxide with increasing hydroxyl ion concentration, the predominant ionic species of polonium(IV) in basic solutions seems to be the $PoO_3{}^{2-}$ ion. As acidity is increased, there is probably a shift to species of the type $PoO_2 \cdot xH_2O$, $Po(OH)_3^+$ and PoO^{2+}, and then to various anionic complexes, for example, $Po(NO_3)_6{}^{2-}$ and $PoCl_6{}^{2-}$ (Wahl and Bonner, 1951).

Although the degree of hydrolysis does not seem to change with perchloric acid concentration in the absence of complexing agent, the change is very rapid in 0.01 mol L^{-1} hydrochloric acid. The following equilibria have been postulated to explain the observations.

$$PoO^{2+} + H_2O \rightleftharpoons PoOOH^+ + H^+ \tag{2.2}$$

$$PoO^{2+} + 2H_2O \rightleftharpoons PoO_3^{2-} + 4H^+ \tag{2.3}$$

$$Po^{2+} + H_2O \rightleftharpoons PoOH^+ + H^+ \tag{2.4}$$

Hydrogen polonide (H_2Po) has been prepared and investigated at a tracer scale (in poor yield) by the action of 2 mol L^{-1} hydrochloric acid on magnesium foil upon which polonium had been deposited chemically or electrolytically, and also by the addition of magnesium powder to a solution of tracer polonium in dilute hydrochloric acid. A stream of dry nitrogen was passed through the acid during the experiment and the emerging gas was scrubbed in aqueous solutions of silver nitrate, lead nitrate, and 2 mol L^{-1} hydrochloric acid. No significant quantities of polonium were carried over in the gas stream. The hydride is not formed from the elements on a milligram scale, is very volatile (boiling point 37° C), and more unstable than bismuth trihydride. It is decomposed by moist air or moist hydrogen (as is hydrogen telluride), by desiccants such as calcium chloride and phosphorous pentoxide, by passage through alkaline solutions and silver nitrate solution and by condensation at low temperatures (Wahl and Bonner, 1951; Bagnall, 1957a; Bagnall et al., 1958a).

2.5 Polonium halides

2.5.1 Physical properties

The behavior of polonium with the halides is somewhat similar to that of tellurium. Polonium forms both a solid dichloride and tetrachloride and divalent anions of both oxidation states. The anion of tetravalent polonium is known to form solid phases with the stoichiometry $M_2(PoX_6)$ (with M = Cs, Rb, K and X = Cl, Br, I). Polonium tetrachloride liquefies at about 300°C forming a pale yellow melt (Abakumov, 1982) that boils at around 390°C. The dichloride is more volatile than the tetrachloride and is known to sublime at about 190°C (Bagnall, 1957b).

Polonium dichloride has been found to have an orthorhombic crystal structure, with the unit cell apparently containing only a single molecule of the salt and, as such, it has been suggested that it is a pseudo-cell (Bagnall, 1957b). There is relatively poor agreement in relation to the lattice parameters (Bagnall et al., 1955a; Moyer, 1956) and the density of the solid has been calculated to be 6.50 g cm^{-3}. Polonium tetrachloride has a monoclinic or triclinic structure (Bagnall, 1957b).

A number of hexachloropolonites have been prepared as solid phases including cesium, ammonium, and tetramethyl ammonium. All have been shown to have a face-centered cubic structure. The cesium phase has a lattice parameter of $a = 10.59$ Å and a density of 3.82 g cm^{-3}, whereas the ammonium phase has a lattice parameter of $a = 10.33$ Å and a density of 2.76 g cm^{-3} (Bagnall, 1957b).

Polonium dibromide melts at around 275°C and sublimes in a vacuum at 110°C (Bagnall, 1957b). The tetrabromide, which is bright red in color, melts at about 330°C. It also has a face-centered cubic structure with a lattice parameter of 5.60 Å. As with chlorine, the phases $Cs_2PoBr_6(s)$ and $(NH_4)_2PoBr_6(s)$ have been prepared and again have a face-centered cubic structure with lattice parameters of 10.99 and 10.82 Å, respectively, and densities of 4.75 and 3.78 g cm^{-3} (Bagnall et al., 1955b).

Polonium tetraiodide is a volatile black solid. It can be used with cesium iodide in a solution of hydriodic acid to produce the solid phase $Cs_2PoI_6(s)$. This latter solid has a face-centered cubic structure with a lattice parameter of 11.77 Å (Bagnall et al., 1956). It is isostructural with $Cs_2TeI_6(s)$ and polonium can coprecipitate with this phase (Bagnall, 1957b).

2.5.2 Chemical properties

The polonium halides are covalent, volatile, readily hydrolyzed compounds of which those being tetravalent are rather less, and the divalent much more, stable than their tellurium analogs. Complex salts of the form M_2PoX_6 (M = Cs, Rb, K; X = Cl, Br, I) have been prepared from the tetravalent halides and closely resemble the corresponding tellurium derivatives, with which they are isomorphous. Of the alkali metals, cesium gives the least soluble complex salt in each case. Salts containing the ammonium and tetramethyl ammonium ions have also been prepared (Staritzky, 1951; Bagnall, 1957a). Absorption spectral studies of polonium chloride solutions have shown that at least two complexes involving polonium and chloride ions exist in hydrochloric acid depending on concentration (Moyer, 1956). Solvent extraction techniques have been used to study the formation of anionic complexes of the type PoX_3^-(or $PoX_3(H_2O)^-$), where X = Cl, Br, I, in weakly acidic solution, and for the formation of the PoX_4^{2-} anion in more concentrated acid. There is electrochemical evidence which suggests polonium(III) species may be transient intermediates in the oxidation of polonium(II) to polonium(IV) in aqueous hydrochloric acid. However, solid trihalides have not been recorded and are very unlikely to be stable with respect to disproportionation. There is also evidence for the existence of a few polonium(IV) mixed halides (e.g., salmon-pink dichlorodibromide, $PoCl_2Br_2$). Polonium(VI) halides are unknown (Bagnall, 1983).

Polonium tetrachloride is hygroscopic and readily hydrolyzed to a white solid of variable composition which is possibly a mixture of a basic chloride and a hydroxide or hydrated oxide. A similar result was obtained by hydrolysis of a solution of the tetrachloride in boiling water. From tracer experiments, it was considered that this product was an oxychloride (Bagnall et al., 1955a). The tetrachloride is soluble in hydrochloric acid, water, with slow hydrolysis, and thionyl chloride, and moderately soluble in ethyl alcohol, acetone, and other ketones. With 0.1 $mol\ L^{-1}$ nitric acid, it gives a white insoluble solid containing no chlorine (Bagnall et al., 1955a). It is slightly soluble in liquid sulfur dioxide but does not appear to react with this solvent. Its reaction with liquid nitrosyl chloride has also been investigated. Nitrogen analysis suggests that the product has four atoms of nitrogen to one of polonium ($(NO)_2PoCl_6 \cdot 2NOCl$ or $(NO)_2PoCl_6 \cdot N_2O_4$) and the compound is readily decomposed by 0.5 $mol\ L^{-1}$ potassium hydroxide (Bagnall et al., 1958a). The adsorption enthalpy and entropy of the tetrachloride onto silica have been reported as $-9\ kJ\ mol^{-1}$ and $-11\ J\ K\ mol^{-1}$, respectively, over the temperature range 300−1100K (Rudolph and Bächmann, 1979).

Solutions in hydrochloric acid are bright yellow at concentrations as low as $5 \times 10^{-5}\ mol\ L^{-1}$ and the addition of a solution of cesium chloride in ethyl alcohol yields a greenish-yellow precipitate of cesium hexachloropolonite ($Cs_2PoCl_6(s)$). The addition of ammonium or sodium hydroxide to solutions in dilute hydrochloric acid precipitates a buff to pale brown flocculent solid, with a solubility of 75 $\mu g\ L^{-1}$ ^{210}Po in water or excess alkali at ambient temperature. When the suspension is boiled, the precipitate becomes crystalline and yellow-brown, and the solubility increases to 12 mg of Po L^{-1}. The precipitate, which is probably a hydrated oxide, appears to be feebly amphoteric (Bagnall et al., 1955a).

In hydrochloric acid solution, the tetrachloride is rapidly reduced to the pink divalent state by sulfur dioxide or hydrazine in the cold, and by arsenic(III) oxide on warming. Hydroxylamine and oxalic acid have no effect and electrochemical observations when using the latter may, therefore, be due to the formation of a complex oxalate. The reduced solutions oxidize back to the tetravalent state after excess reducing agent is eliminated. The change of potential with time at a platinum electrode was measured to determine the valency state of the reduced polonium. The values are 620 mV at 4 minutes [Po^{2+} (pink)] and 470 mV at 12 minutes [Po^{4+} (yellow)]. The break in the potential−time curve (510 mV at

7.5 minutes) could correspond to the formation of an intermediate trichloride [Po^{3+} (pink)]. However, there is no other evidence for the existence of this compound (Bagnall et al., 1955a).

In coprecipitation experiments, it was found that salts of $PoCl_6^{2-}$ are isomorphous with chloroplumbates, chloroplatinates, and chlorotellurites. Potassium, rubidium, cesium, ammonium, and tetramethylammonium salts are insoluble and $PoCl_6^{2-}$ has been demonstrated as the principal species present in solution in 2.5 mol L^{-1} hydrochloric acid (Cairo, 1958).

Dark, ruby-red polonium dichloride is formed by the reduction of the solid tetrachloride with sulfur dioxide at 25ºC. The solid is hygroscopic and mildly volatile (Bagnall, 1983). The dichloride dissolves readily in dilute hydrochloric acid to form a pink solution that rapidly oxidizes, or is immediately oxidized by hydrogen peroxide or chlorine water, to the tetravalent state. The addition of potassium hydroxide to this solution gives a dark brown precipitate (solubility 1.4 mg of Po L^{-1}), which may be the hydrated divalent oxide or hydroxide, and which is very rapidly oxidized to the tetravalent state (Bagnall et al., 1955a). When the dichloride is heated in ammonia gas at 200ºC, a brown ammine forms (Bagnall, 1983). In 0.1 mol L^{-1} nitric acid, the dichloride forms a dark red solution and then, rapidly, a white flocculent precipitate, the composition of which is not known (Bagnall et al., 1955a). The adsorption enthalpy and entropy of the dichloride onto silica have been studied using gas chromatography and were found to be -133 kJ mol^{-1} and -74 J K mol^{-1}, respectively, over the temperature range 300–1100K (Rudolph and Bächmann, 1979).

Solutions of both polonium tetra- and dichlorides yield black precipitates with hydrogen sulfide. Heating these precipitates in a vacuum gives sulfur and metallic polonium. Although these precipitates may be sulfides, their exact composition has not been determined (Bagnall et al., 1955a).

Polonium tetrachloride is converted to a white solid by 15% hydrobromic acid, which gives polonium dioxide on heating. The solid is probably a product of the hydrolysis of the tetrachloride by the hydrobromic acid rather than a bromate (Bagnall et al., 1958a).

Polonium(IV), together with selenium(IV) and tellurium(IV), are known to form chloride complex anions of the type MCl_5^- or MCl_6^{2-} in concentrated hydrochloric acid. Trace polonium can be separated from a solution of the group 6B elements on Dowex 1X-4 anion exchange resin (chloride form). Polonium is eluted using either nitric or perchloric acids, although the effectiveness of perchloric acid as an eluant is time-dependent (Sasaki, 1955; Bagnall, 1957b).

Anion exchange has also been used in determining the average anionic charge of polonium species in chloride solution. The method was based on measurements of the distribution ratio of polonium at a constant internal chloride ion concentration of the anion exchanger phase. In 1 mol L^{-1} sodium chloride/hydrochloric acid around pH 1, polonium (IV) was found to exist in both chemical forms $PoCl_4(OH)_2^{2-}$ (80%) and $PoCl_3(OH)_2^{-}$ (20%) (Suganuma, 1995).

Polonium tetrabromide is hygroscopic and easily hydrolyzed, yielding a white solid of indefinite composition, presumably a basic bromide. It is soluble in hydrobromic acid, ethyl alcohol, acetone, and some other ketones, sparingly soluble in liquid bromine and insoluble in benzene, chloroform, and carbon tetrachloride. The tetrabromide dissolves in dilute hydrobromic acid to give an orange-red (10^{-3} mol L^{-1}) or carmine-red (0.025 mol L^{-1}) solution. Addition of aqueous cesium bromide to a 10^{-3} mol L^{-1} solution yields an immediate precipitate of dark red cesium hexabromopolonite ($Cs_2PoBr_6(s)$), which is immediately hydrolyzed by cold water. It is rapidly reduced in solution to the pink divalent state by sulfur dioxide or hydrazine in the cold. Reduced solutions are reoxidized to the tetravalent state, and the oxidation potential—time curve shows no evidence for the existence of an intermediate trivalent bromide. With ammonia at room temperature, the tetrabromide forms an unstable yellow ammine and gives some indication of a volatile, colorless phase (Bagnall et al., 1955b).

Polonium dibromide is a purple-brown solid, formed by the reduction of the solid tetrabromide with sulfur dioxide at 25°C. The reduction, however, is incomplete. By analogy to the dichloride, the solid is hygroscopic and somewhat volatile (Bagnall, 1983). It is soluble in a number of ketones and dilute hydrobromic acid, giving purple solutions that are rapidly oxidized to the tetravalent state. Solutions of the dibromide in hydrobromic acid are obtained by similar methods to those of the dichloride (i.e., reducing the tetrabromide in hydrobromic acid with sulfur dioxide or hydrazine in the cold or with arsenic(III) oxide on warming) (Bagnall et al., 1955a). In dilute nitric acid, the dibromide decomposes to form a white precipitate of unknown composition (Bagnall et al., 1955b).

Polonium tetraiodide is the only iodide known and is prepared by treating the hydroxide or dioxide with 0.1 mol L^{-1} hydroiodic acid or by precipitation from a hydrochloric acid solution of the tetrachloride with 0.1 mol L^{-1} hydroiodic acid. It is insoluble in 2 mol L^{-1} hydrochloric acid, 1 or 2 mol L^{-1} nitric acid, acetic acid, chloroform, benzene, carbon tetrachloride, and diethyl and dibutyl ether, and is slightly soluble in ethyl alcohol and acetone. It is slowly hydrolyzed to a white solid of indefinite composition in water and is decomposed by hot, concentrated nitric acid, or sodium hypochlorite (NaOCl), or slowly by concentrated potassium hydroxide. It is soluble in 2 mol L^{-1} hydriodic acid, giving a solution that is red-brown at 20°C and green at 0°C (Bagnall et al., 1956). The tetraiodide is oxidized by acidified potassium nitrite and other oxidizing agents (Bagnall, 1957a). A suspension of the solid in 0.1 mol L^{-1} hydriodic acid is not reduced by sulfur dioxide or hydrazine, and no precipitate is obtained when dilute solutions of hydriodic acid or potassium iodide are added to solutions of polonium dichloride in hydrochloric acid (Bagnall et al., 1956). Solid tetraiodide, however, is reduced to the metal by hydrogen sulfide (Bagnall, 1983).

The addition of a solution of cesium iodide in 0.1 mol L^{-1} hydroiodic acid yields an immediate precipitate of black cesium hexaiodopolonite ($Cs_2PoI_6(s)$), which is readily hydrolyzed by water. The corresponding potassium and rubidium salts are insoluble. Solubility studies of the tetraiodide in 0.02–0.5 mol L^{-1} hydroiodic acid, carried out from 0°C to 50°C, indicate that the solubility is proportional to the square of the acid concentration. For a fixed concentration of 0.3 mol L^{-1} hydroiodic acid, the relationship of solubility to acid concentration at 0°C, 22°C, and 50°C was found to be identical, and therefore, the same complex ion must be involved (Bagnall et al., 1956).

In solutions containing hydroiodic acid and lithium iodide of constant total iodide ion concentration and varying hydrogen ion concentration, the results show that the solubility is independent of the hydrogen ion concentration, and therefore, the reaction involved must be that shown by Eq. (2.5).

$$PoI_4(s) + 2I^- \rightleftharpoons PoI_6^{2-} \tag{2.5}$$

The equilibrium constant, *K*, determined was $5.9 \pm 0.2 \times 10^{-3}$ at 22°C. This relationship, however, does not hold for (HI) < 0.02 mol L^{-1} and it is possible that the deviations at low acid concentrations are due to reaction (2.6).

$$PoI_4(s) + I^- \rightleftharpoons PoI_5^- \tag{2.6}$$

The equilibrium constant for reaction (2.6) can be estimated by calculating the solubility due to reaction (2.5) and subtracting this result from the observed value. The value obtained for *K* is $6.7 \pm 0.5 \cdot 10^{-5}$ at 22°C.

Solvent extraction data obtained with tracer level polonium also indicate the formation of the halocomplex anions PoX_5^-(or $PoX_5(H_2O)^-$) in dilute halogen acid solution and the hexahaloanions, PoX_6^{2-} in more concentrated acid solutions (X = Cl, Br, I). The equilibrium constant for the formation of the $PoCl_6^{2-}$ ion (reaction 2.7) is ~10^{14}, as determined from electrochemical data (Bagnall, 1983).

$$Po^{4+} + 6Cl^- \rightleftharpoons PoCl_6^{2-} \tag{2.7}$$

A white solid, presumably polonium tetrafluoride or a basic compound, results from the action of dilute aqueous hydrofluoric acid on polonium hydroxide or tetrachloride. On treatment with sulfur dioxide, the solid becomes bluish gray (possibly due to reduction to the divalent state), while on standing, the material reverts to the white solid. The solubility of tetravalent polonium in aqueous hydrofluoric acid increases rapidly with increasing concentration, indicating complex ion formation (Bagnall et al., 1958a). The formation of a volatile polonium fluoride compound that subsequently decomposes due to chemical or radioactive decomposition has been described by Weinstock and Chernick (1960). The formation of a difluoride has not been recorded (Bagnall, 1983).

Moyer (1956) summarized studies on the reaction of halogen vapors with pure polonium metal. The metal is heated in dry chlorine at 1 atm from 125°C to 200°C and the color changes progressively from gray to brown to yellow. The yellow tetrachloride, $PoCl_4$, volatilizes at about 390°C in a chlorine atmosphere. At lower temperatures and pressures, a red compound forms, which was shown to be polonium dichloride, $PoCl_2$. When the metal is treated with dry bromine vapor at 200 mm Hg, a reaction occurs on standing overnight at room temperature. This reaction is then completed by heating for 1 hour at 250°C and, after volatilizing at 360°C in a bromine vapor atmosphere, dark red crystals of polonium tetrabromide ($PoBr_4$) condense. No conclusive evidence for the formation of polonium tetraiodide (PoI_4) was obtained and no volatile fluorides were formed when the metal was treated with fluorine from room temperature to 700°C.

MIBK and acetylacetone extracted all the polonium from the aqueous phase of halogen acid solutions over a wide range of acid concentrations. Evaporation of the extracts left a yellow oil, which is miscible with 60/80 petroleum ether on warming. With strong cooling, pale yellow needles separate from the solution and the polonium can only be recovered from this crystalline product by destroying the organic component with a strong oxidizing agent (Bagnall et al., 1958a). The reaction of polonium tetrachloride or tetrabromide with MIBK also yields dihalopolonium(IV) compounds, possibly of the form $(iC_4H_9COCH_2)_2PoX_2$ (Bagnall, 1983).

There is very little published information on complexes of the tetrahalides with neutral donor ligands. Amines of unknown composition are formed when polonium tetrachloride or tetrabromide are exposed to gaseous ammonia and a study of the solubility of the tetrachloride in tributyl phosphate (TBP) indicates the formation of the complex $(PoCl_4(TBP)_2)$. This area of polonium chemistry, however, requires further investigation (Bagnall, 1983).

2.6 Polonides, polonites, and the polonium compounds with other chalcogens

2.6.1 Physical properties

A relatively large number of polonide phases have been prepared and their crystal structures determined. Lattice parameters for sodium polonide were predicted on the basis of those for Na_2O, Na_2S, Na_2Se, and Na_2Te and the parameters that were determined following the preparation of Na_2Po were in good agreement with the prediction. The phase has a face-centered cubic structure (fluorite type) with a lattice parameter of $a = 7.473$ Å. Sodium polonide has a calculated density of 4.08 g cm^{-3} (Moyer, 1956). It has also been demonstrated that Na_2Po is isomorphous with Na_2Te (Khlopin and Samartseva, 1934).

Beryllium, magnesium, calcium, strontium, and barium polonide phases have all been prepared by Witteman et al. (1960). Beryllium polonide has a face-centered cubic structure of the sphalerite type with a lattice parameter of $a = 5.838$ Å and a density of 7.3 g cm^{-3} (Witteman et al., 1960). It is volatile at temperatures above 600°C. Unlike the other alkaline earth polonides, magnesium polonide has a hexagonal structure of the NiAs type. It has lattice parameters of $a = 4.345$ and $c = 7.077$ Å and a calculated density of 6.7 g cm^{-3}. Calcium, strontium, and barium polonide all have face-centered cubic structures of the halite type. They have lattice parameters and densities of $a = 6.514$ Å and 6.04 g cm^{-3}, $a = 6.796$ Å and 6.3 g cm^{-3}, and $a = 7.119$ Å and 6.3 g cm^{-3}, respectively. These latter phases are volatile above about 600°C–650°C (Witteman et al., 1960).

The polonide phases of all the rare earth elements (Sc, Y, and the lanthanides) have been studied. Two phases have been reported for each rare earth metal [MPo(s) and M_2Po_3(s)]. The latter phases are obtained by heating the rare earth metals in polonium vapor for a period up to about 11 hours at 1000°C. The monopolonide phases are obtained at lower temperatures in a shorter period (up to 2 hours). The melting point of the M_2Po_3(s) phases range between 1442°C and >2400°C whereas those of the monopolonide phases are slightly lower at 1235°C–2212°C. More detail on these phases is described by Abakumov (1982).

Nickel polonide has been produced when nickel is heated in polonium vapor. Although its stoichiometry and lattice parameters are variable, it has been reported to have a hexagonal structure (Moyer, 1956; Abakumov, 1982) and a melting point near 625°C (Witteman et al., 1960).

Polonide phases have been reported for zinc, cadmium, and mercury. As might be expected, zinc polonide has the sphalerite structure with a lattice parameter of $a = 6.309$ Å and a density of 7.2 g cm^{-3}. Cadmium polonide (CdPo) also has the sphalerite structure with a lattice parameter of $a = 6.665$ Å and a density of 7.2 g cm^{-3}. Mercury polonide differs in its structure being of the halite type. It has a lattice parameter of $a = 6.25$ Å and a density of 11.1 g cm^{-3}. These properties were reported by Witteman et al. (1960).

Other polonide phases whose physical properties have been studied are those of silver, platinum, lead, and bismuth. Silver polonide was prepared as early as 1950 (Moyer, 1956). An orthorhombic structure was assigned but this was not entirely satisfactory and it was suggested that the structure may be either orthorhombic or monoclinic, which correspond to the structures of silver selenide and silver telluride, respectively. Lead polonide has the halite structure with a lattice parameter of $a = 6.59$ Å and a density of 9.6 g cm^{-3} (Witteman et al., 1960). The vapor pressure of lead polonide has also been measured and the vaporization enthalpy has been reported to be 139 kJ mol^{-1}. Platinum polonide is believed to have the stoichiometry $PtPo_2$(s) (Moyer, 1956). It is thought to have a hexagonal structure of the $Cd(OH)_2$ type. Bismuth polonide has been assigned two structures; a rhombohedral ($a = 4.456$–4.503 Å; $c = 3.602$ Å) and a cubic structure ($a = 3.602$ Å), but no definitive structure has been assigned (Abakumov, 1982).

Studies of other polonide phases have been undertaken, but no definitive data have been reported. Discussion of these phases is provided in the review of Abakumov (1982).

2.6.2 Chemical properties

Polonium monosulfide, PoS, forms as a black precipitate by the action of hydrogen sulfide on solutions of polonium di- or tetrachloride in dilute hydrochloric acid. It is soluble in concentrated hydrochloric acid, insoluble in ethyl alcohol, acetone, and toluene and decomposed by bromine, sodium hypochlorite, aqua regia, and ammonium sulfide (Bagnall, 1957a; Bagnall and Robertson, 1957b). The sulfide also forms by treating polonium(IV) hydroxide with aqueous ammonium sulfide and is decomposed to the elements at 275°C under vacuum (Bagnall, 1983).

The solubility product was determined by precipitating the compound from solutions of varying hydrochloric acid concentration which had been saturated with hydrogen sulfide. The sulfide ion concentration was calculated from the solubility data of Kendall and Andrews (1921) and the known dissociation constants of hydrogen sulfide, but activity corrections were not applied. The reproducibility was not good and attainment of equilibrium required some time, during which a considerable amount of sulfur was precipitated, probably due to the oxidation of hydrogen sulfide by alpha radiation. The solubility product was found to be $5.5 \pm 0.1 \times 10^{-29}$ in $1-5.5\ \mathrm{mol\ L^{-1}}$ hydrochloric acid at ambient temperature (Bagnall and Robertson, 1957b).

The formation of polonium(IV) sulfate provides further evidence for the more markedly basic character of polonium dioxide as compared with tellurium dioxide, which is to be expected from its position in the Periodic Table. The disulfate is of particular interest since no tellurium analog is known. The white, hydrated disulfate, $Po(SO_4)_2 \cdot xH_2O$, is obtained when polonium tetrachloride or the hydrated dioxide is treated with sulfuric acid ($>0.25\ \mathrm{mol\ L^{-1}}$). Removal of the supernatant leads to a series of irreversible color changes at room temperature. These are also observed on heating (pink at 200°C, deep purple at 380°C) and are probably due to progressive dehydration. The solid decomposes to polonium dioxide at 550°C. The disulfate phases are very soluble in dilute hydrochloric acid and insoluble in acetone and ethyl alcohol. The solubility in dilute sulfuric acid is remarkably low (420 μg of anhydrous salt $\mathrm{L^{-1}}$ in $0.25\ \mathrm{mol\ L^{-1}}$ sulfuric acid; no temperature given); the observed increase in solubility with acid concentration suggests complex formation. The solid loses water at 100°C and when washed with anhydrous diethylether, yielding the deep purple, anhydrous disulfate (Bagnall and Freeman, 1956b; Bagnall, 1957a, 1983).

Studies of diffusion, electrochemical, and solvent extraction behavior of polonium as a tracer in sulfuric acid appear to indicate the formation of complex ions, although there are no data on their composition. Tracer polonium in sulfuric acid solution has been reduced by hydroxylamine at the boiling point. The reaction involved and its redox potential are not known. Work with milligram amounts of polonium confirms this, since suspensions of the disulfate in 0.5−1 mol L^{-1} sulfuric acid dissolve when boiled with hydroxylamine, yielding a pink solution. This color appears to be characteristic of divalent polonium sulfate. The tetravalent sulfate, however, reprecipitates after cooling for some minutes, even in the presence of excess hydroxylamine (Bagnall and Freeman, 1956b; Bagnall, 1957a).

In dilute sulfuric acid (0.05−0.25 mol L^{-1}), a white, basic sulfate of composition $2PoO_2 \cdot SO_3$ is formed. This compound is analogous to the only known tellurium(IV) sulfate, $2TeO_2 \cdot SO_3$. It can also be formed by hydrolysis of the disulfate and is yellow above 250°C. It is more soluble in dilute sulfuric acid than the disulfate and is readily soluble in dilute hydrochloric acid. Solubility studies indicate that it is metastable with respect to the disulfate within this acid range (0.05−0.25 mol L^{-1}). It also decomposes to polonium dioxide at 550°C (Bagnall and Freeman, 1956b; Bagnall, 1957a, 1983).

The basic sulfate appears to be metastable, since the solubility curve of the disulfate can be extended to regions of lower acid concentration by diluting the acid in contact with solid polonium disulfate to concentrations at which the basic sulfate is normally formed. Further, seeding the aqueous phase in contact with the basic sulfate with small crystals of the disulfate decreases the solubility to a marked degree. Determinations made at sulfuric acid concentrations between 0.15 and 0.25 mol L^{-1} give erratic results. The increase in solubility with acid concentration appears to indicate complex formation, although no simple relationship has been deduced (Bagnall and Freeman, 1956b).

Basic polonium selenate, $2PoO_2 \cdot SeO_3$, is a white solid formed by treating polonium tetrachloride or hydroxide with 0.007−2.5 mol L^{-1} selenic acid. It is very soluble in dilute hydrochloric acid and is deep yellow above 250°C (Bagnall and Freeman, 1956b). The solubility in selenic acid increases 100-fold with increasing acid concentration from 0.025 to 2.5 mol L^{-1} (Bagnall, 1983). The solubility curve for the basic selenate in selenic acid does not show the discontinuity found for the analogous sulfate compound. It is possible that the diselenate is formed at higher acidities; however, attempts to prepare the compound by evaporation of solutions of the basic selenate in concentrated selenic acid were unsuccessful (Bagnall and Freeman, 1956b).

Tracer experiments suggest sodium polonide (Na_2Po) results from the reduction of polonium by sodium dithionite in alkaline solution under an atmosphere of hydrogen (Bagnall, 1957a). Reaction of polonium metal vapor with mixtures of metals in a melt usually yields a mixture of the appropriate polonides. Polonides produced in this way are analogous to the corresponding chalcogenides, H_2X ($X = S$, Se, Te), and in many cases these compounds can be regarded as being derived from H_2Po. In most cases, the metal polonides are isostructural with the analogous tellurides, although some, HgPo (NaCl type) and HgTe (α-ZnS type), indicate that the polonide structure is more ionic in character than that of the telluride. Polonium vapor does not react with silicon, chromium, zirconium, or tantalum carbides, or ruthenium, osmium, rhenium, technetium, or cobalt. Reaction to form alloys or polonides occurs with many other elements at temperatures between 400°C and 1000°C (Bagnall, 1983).

Most polonides dissociate to the elements at moderately high temperatures and a mass spectrometric study has identified Po^+ in the decomposition of lanthanum polonide (LaPo), neodymium polonide (NdPo), dysprosium polonide (DyPo), and gadolinium polonide (GdPo). From vapor pressure studies of the dissociation of gadolinium polonide and indium polonide (InPo), both of which are liquid, there is evidence for the dissociation shown in Eq. (2.8).

$$3MPo \rightleftharpoons M_3Po_2 + Po(g) \tag{2.8}$$

Calculated heat capacities at constant pressure and standard entropies and enthalpies have been reported for carbon polonide (CPo), germanium polonide (GePo), and tin polonide (SnPo), the last of which seems to be reasonably well established as a compound. All known polonides are rapidly oxidized in air [e.g., titanium polonide (TiPo), zirconium polonide (ZrPo), hafnium polonide (HfPo)], probably due to alpha radiation effects (Bagnall, 1983).

2.7 Polonium nitrates

2.7.1 Chemical properties

Experiments have shown that solid polonium hydroxide or tetrachloride reacts immediately with dilute ($0.1-2\ mol\ L^{-1}$) nitric acid to give a white, crystalline solid, which readily decomposes to polonium dioxide under vacuum or on heating and is easily hydrolyzed to the tetravalent hydroxide. Attempts to analyze the solid have failed and it is most likely an unstable addition compound like $PoO_2 \cdot xHNO_3$ (Bagnall et al., 1958b).

On long-standing in 0.5 mol L^{-1} nitric acid (12 hours), both the hydroxide and tetrachloride give a white crystalline product which can be dried under vacuum at room temperature with only slight decomposition. This product also forms when polonium metal is left in contact with a mixture of nitrogen dioxide and oxygen. The product reacts readily with water, dilute aqueous potassium hydroxide, or dilute hydrochloric acid to give the tetravalent hydroxide or chloride, and yields polonium metal on long-standing in a stream of dry nitrogen or under vacuum. The NO_3^-: Po ratio is 3:2 at temperatures above 100°C and 1:2 below 100°C (Bagnall et al., 1958b).

When a solution of polonium hydroxide in 2 M nitric acid is evaporated and dried under vacuum, a yellowish-white powder results, which has a NO_3^-:Po ratio of 1:2, analogous to the only known tellurium nitrate. The product decomposes to polonium dioxide at ~130°C and also yields polonium metal on long-standing in a stream of dry nitrogen or under vacuum (Bagnall et al., 1958b).

The reaction of polonium dioxide and tetrachloride with liquid dinitrogen tetroxide (N_2O_4) results in the formation of the white, solvated tetranitrate, $Po(NO_3)_4 \cdot (1 + x)N_2O_4$. The solid is insoluble in liquid dinitrogen tetroxide, very easily hydrolyzed and, under vacuum, decomposes to a white, basic nitrate with a NO_3^-:Po ratio of 3:2. The solid behaves similarly to the two products described above and it has been suggested that the structures are probably oxygen-bridged dimers or polymers (Bagnall et al., 1958b; Bagnall, 1983).

White, crystalline compounds, presumably the nitrate and nitrite, have also been obtained by the action of aqueous 1 mol L^{-1} potassium nitrate or nitrite on solid polonium tetrachloride. Both solids convert to the tetravalent hydroxide on standing and in contact with excess reagent (Bagnall et al., 1958b).

Polonium metal is vigorously attacked by concentrated nitric acid, giving a yellow solution when at 10^{-5} mol L^{-1} with respect to polonium. Evaporation yields a white solid of uncertain composition. The solubility of these polonium nitrates in nitric acid has been determined over a wide range of concentrations and temperatures. The observed solubilities are extremely low (the same order as barium sulfate in water), and therefore, it seems probable that a basic nitrate is involved. Solubility and cation exchange studies indicate that anionic nitrato-species are formed at high ($>2\ mol\ L^{-1}$) concentrations of nitric acid. Although definitive information is not available, complex ions such as $Po(NO_3)_5^-$ may be present (Bagnall, 1957a). Salts of these complex anions have not been recorded (Bagnall, 1983).

The solubility of polonium in 0.1–8 mol L^{-1} nitric acid at temperatures of 25°C, 35°C, and 45°C was studied by Moyer (1956). The quantity of polonium dissolved varied from 4.0 mg L^{-1} at the lowest temperature and acid concentration to 970 mg L^{-1} at the highest temperature and acid concentration. There was an inflection point on the solubility curve between 1.1 and 1.2 mol L^{-1} acid, and reactions (2.9) and (2.10) have been proposed for the reactions in <1.1 and in >1.2 mol L^{-1}, respectively.

$$PoO(NO_3)_2(s) + HNO_3 \rightleftharpoons PoO(NO_3)_3^- + H^+ \tag{2.9}$$

$$Po(NO_3)_4(s) + HNO_3 \rightleftharpoons Po(NO_3)_5^- + H^+ \tag{2.10}$$

Alternative explanations of this inflection point are provided in Chapter 3, Chemical thermodynamics of polonium.

2.8 Solvent extraction of polonium

Solutions of TBP in organic solvents can be used to separate polonium from solutions containing lead and bismuth. Extraction is strongly dependent on the acid concentration and reaches a maximum for solutions between 6 and 9 mol L^{-1} hydrochloric acid (Bagnall, 1957a; Schulz et al., 1987; Lally, 1992). Solubility studies suggest that the complex formed is $PoCl_4 \cdot 2TBP$, which has an equilibrium constant, $K = (PoCl_4 \cdot 2TBP)/(TBP)^2$, of about 0.04 at 22°C (Bagnall and Roberston, 1957a). Polonium has also been extracted from solutions containing sulfuric acid and ferrous sulfate (Bagnall, 1957b).

Dithizone dissolved in chloroform removed traces of polonium from both hydrochloric and nitric acid solutions at pH 0.2–5. It is probable that the complex extracted by both acids is $PoODz_2$ (Dz = dithizonate ion); however, the volatility of the complex made it hard to determine the polonium content with any accuracy (Bagnall, 1957a). In a solvent extraction study, Bagnall and Roberston (1957a) used the dependence of the partition coefficient on the acid concentration of the aqueous phase and also found $PoODz_2$ to be the likely formula for the complex, since polonium remained in the tetravalent state under the conditions of the experiment. Dithizone has also been used to extract polonium from ammoniacal potassium cyanide and ammonium citrate solutions (Bagnall, 1957b).

Suganuma and Hataye (1981) investigated the hydrolysis and chlorocomplex formation of polonium(IV) using dithizone-carbon tetrachloride solutions. In 1.0 mol L^{-1} (H, Na)Cl, polonium was found to exist as $Po(OH)_2Cl_4^{2-}$, $Po(OH)_2Cl_{3.7}^{1.7-}$, $Po(OH)_2Cl_{3.5}^{1.5-}$, $Po(OH)_{2.1}Cl_{2.9}^{-}$, $Po(OH)_4$, at pH 0, 1, 2, 3, and 6, respectively. The chemical compositions of the extracted species were probably $PoCl_2(HD_z)_2$, $Po(OH)Cl(HD_z)_2$, or $Po(OH)_2(HDz)_2$. The logarithms of the successive hydrolysis constants were calculated from the distribution ratios obtained and were found to be 4.6 ± 0.3 and 8.7 ± 0.3, respectively (ambient temperature). Hataye et al. (1981) studied the hydrolysis of polonium(IV) in perchlorate solutions using similar conditions and obtained a log K of 1.12 in 1.0 mol L^{-1} (H, Na)ClO_4 at room temperature. Starik et al. (1964a) investigated the hydrolysis of polonium(IV) in perchloric acid solutions using solvent extraction. The formation constants for the following hydroxy species, $PoOH^{3+}$, $Po(OH)_2^{2+}$, $Po(OH)_3^{+}$, and $Po(OH)_4$ were 5 × 10^{12}, 2.5 × 10^{25}, 2.2 × 10^{38}, and 2.5 × 10^{50}, respectively, at 18°C–22°C. Ampelogova (1973) reanalyzed this work and found evidence for the species PoO^{2+}, $PoO(OH)^{+}$, and $PoO(OH)_2$(aq) only.

Ampelogova (1973) studied the complexation of polonium(IV) by sulfate using solvent extraction. The stability constants obtained for $PoOSO_4$ and $PoO(SO_4)_2^{2-}$ in 2 mol L^{-1} H(ClO_4, HSO_4) were 29 and 2500, respectively. He also studied the complexation of polonium(IV) by nitrate and obtained stability constants for $PoONO_3^{+}$, $PoO(NO_3)_2$(aq) and $PoO(NO_3)_3^{-}$ of 0.56, 1.15, and 1.30 in 1 mol L^{-1} and 0.53, 1.08, and 1.30 in 1.5 mol L^{-1} nitric acid, respectively (ambient temperature).

Polonium has been extracted from bismuth using 0.25 mol L^{-1} thenoyltrifluoroacetate in benzene from aqueous solution between pH 0 and 2 (Bagnall, 1957a,b; Lally, 1992). Mesityl oxide has been used to quantitatively extract polonium, present in concentrations below 10^{-9} g mL^{-1}, from nitric acid solution saturated with aluminum nitrate (Marechal-Cornil and Picciotto, 1953).

Acetylacetone and MIBK remove milligram amounts of polonium from dilute hydrochloric acid almost quantitatively. The extractions probably depend on condensation with the ketones, analogous to the behavior displayed by tellurium tetrachloride (Bagnall, 1957a). Cairo (1958) studied the extractability of polonium with diisopropylketone and proposed structures of H_2PoCl_6, $H_2Po(NO_3)_6$ and $H_2Po(SO_4)_3$ for the extracted species in hydrochloric, nitric, and sulfuric acids, respectively.

Polonium diethylammonium diethyldithiocarbamate (DDTC) was formed in acidic solution and extracted with chloroform, carbon tetrachloride, and amyl alcohol. The complex had one DDTC anion for every polonium atom (Kimura and Ishimori, 1958). Wai and Lo (1982) and Lo and Wai (1983) separated ^{210}Po and ^{210}Bi from ^{210}Pb in nitric acid solution using diethyldithiocarbamate (DTC). From the distribution ratio measured at equilibrium, the extraction constant for $Po(DTC)_4$ was estimated to be 2.75×10^{56} at ambient temperature.

Polonium in aqueous solution can be extracted with 8-hydroxyquinoline in chloroform, probably due to the formation of polonium hydroxyquinolate, with one hydroxyquinolate anion for every polonium atom. It can also be extracted from acetic acid solution using chloroform, giving 100% extraction at pH 4.2. The thionalide salt has two organic anions for one atom of polonium. It is not extracted from hydrochloric or hydrobromic acids with diethyl- or diisopropylethers. It is, however, extracted from hydrochloric acid solutions containing potassium iodide with isopropylether and MIBK (Kimura and Ishimori, 1958).

Polonium-210 is extracted by 5% tri-*n*-benzylamine in chloroform from 6 mol L^{-1} hydrochloric acid (>99%), 6 mol L^{-1} nitric acid (<10%) and 5 mol L^{-1} hydrochloric/0.5 mol L^{-1} nitric acid (93%) using an organic to aqueous ratio of 1.7:1 (Moore, 1957). Cupferron in a solution of amyl acetate extracts trace polonium from 3 mol L^{-1} mineral acid in the presence of sulfurous acid (Bagnall, 1957b). Polonium is extracted from concentrated nitric acid solution using diethylether containing peroxide (Danon and Zamith, 1957).

2.9 Polonium ion exchange

The state of polonium in aqueous solution has been investigated using ion exchange equilibrium studies. The distribution of polonium between the solid and liquid phases was determined for solutions of various compositions using both a cation [Dowex-50 (H^+ form)] and an anion (Dowex-2) exchanger. The studies showed that in the absence of a complexing agent, polonium is hydrolyzed to such a degree that it forms monovalent cations and anions. On the basis of these results, and of previous work, it appears that these ions may be $PoO(OH)^+$ and $HPoO_3^-$. The addition of a complexing anion, such as chloride, decreases both the cationic and anionic form of polonium, up to a chloride concentration of about $0.01\ mol\ L^{-1}$. At higher chloride concentrations, the cation concentration continues to decrease while the percentage of polonium in the anionic state increases. In the presence of chloride, the hydrogen ion concentration is also very important in determining the fractions of polonium in the cationic, neutral, and anionic states (Tompkins, 1951).

Tompkins postulated that polonium can exist in aqueous solutions as either a cation or an anion. It is also probable that it is present as neutral complexes, particularly in the presence of complexing agents, since it can be extracted into nonpolar liquids. Because of the tendency of polonium to form coordinate bonds, it does not exist in aqueous solutions as simple ions. It forms complexes with anions such as nitrate, chloride, and sulfate, and in their absence, or in dilute solutions, is hydrolyzed. Depending upon the degree of hydrolysis, the following species have been postulated:

Po(IV) $PoOH^{3+}, Po(OH)_2^{2+}, PoO_2^+,$
$PoOOH^+, Po(OH)_3^+, PoO(OH)_2(aq), PoO_2(s), PoO_2OH^-, PoO_3^{2-}$; and

Po(II) $PoOH^+, Po(OH)_2(aq), PoO(s), PoOOH^-, PoO_2^{2-}$.

By determining the equilibrium distribution of polonium between cation and anion exchangers and perchlorate solutions of various concentrations, information about the cations and anions should be obtained. If the hydrogen ion concentration is varied, information about the hydrolysis may also be revealed.

Tompkins' results indicate that polonium exists as monovalent cations in $4-8\ mol\ L^{-1}$ perchloric acid solutions. The much lower distribution coefficients using an anion exchange resin indicate that the polonium is largely cationic. When $4\ mol\ L^{-1}$ perchloric acid is used as the bulk electrolyte and the chloride concentration varied from 0.0001 to $0.1\ mol\ L^{-1}$, the percentage of the cationic form is found to decrease rapidly as the concentration of chloride is increased up to $\sim 0.01\ mol\ L^{-1}$. A slight decrease is observed in the anionic percentage. This result indicates that a neutrally charged complex is most stable under these conditions, being formed from both cationic and anionic polonium. As the chloride concentration is further increased, the distribution coefficient of polonium on the resin increased. From ~ 0.008 to $0.1\ mol\ L^{-1}$, this increase appears to have a first power dependence on the chloride concentration. A further experiment, in which the ionic strength and hydrogen ion concentration were kept constant at 1 mol L^{-1} acid by varying the ratio of perchloric to hydrochloric acids in the solution, indicated that the chloride ion forms both mono- and divalent anions with polonium over the chloride concentration range 0.03–0.7 mol L^{-1}.

Tompkins also performed a series of experiments to determine the effect of varying the perchloric acid concentration (1.0–8.0 mol L^{-1}) at constant chloride concentration (0.01 mol L^{-1}). The cations in solution decrease rapidly as the acid concentration is increased. The data indicate that up to ~4 mol L^{-1} perchloric acid, a neutral complex is formed. Above this concentration, the rate of decrease of cations remains constant, while the anion concentration increases rapidly. On the basis of the experiments, it is evident that both polonium(IV) and polonium(II) are highly hydrolyzed in aqueous solutions. Both ions form monovalent cations in perchloric acid solutions, though only the higher valence state appears to form anions. As the chloride concentration is increased in the presence of 4 mol L^{-1} perchloric acid, the oxide and hydroxide groups seem to be replaced by chloride to form, first, another cation, then a neutrally charged complex and finally one or more anions.

Although the degree of hydrolysis does not seem to change with perchloric acid concentration in the absence of complexing agent, the change is very rapid in 0.01 mol L^{-1} hydrochloric acid. The oxygen and hydroxide groups appear to be successively replaced by chloride to form, first, a neutrally charged complex and finally, in higher perchloric acid concentrations, anions.

Ishimori (1955), Casella et al. (1982), and Lally (1992) have reported polonium adsorption onto strongly basic anion exchange resins from hydrochloric acid solution. Suganuma (1995) used anion exchange to determine the average ionic charge of polonium species in chloride solutions. Marsh et al. (1978b) studied the anion exchange behavior of polonium in 0.1–8.7 mol L^{-1} hydrobromic and 0.1–7.4 mol L^{-1} hydroiodic acids. Tracer levels of polonium were completely adsorbed onto strong based anionic resins from all concentrations of both acids.

Nelson et al. (1964) studied the cation exchange behavior of polonium in hydrochloric and perchloric acid solutions. They found that polonium was negligibly adsorbed from 9 mol L^{-1} concentrations of both acids and that adsorbability was also low ($D_v < 1.5$) from more dilute solutions (0.2–1 mol L^{-1} HCl or $HClO_4$). Bagnall (1957b) used cation exchange to separate polonium from bismuth in 0.1–0.3 mol L^{-1} hydrochloric acid solutions. Bagnall et al. (1958b) and Marsh et al. (1978a) studied the cation exchange behavior of polonium in nitric acid. Polonium adsorption onto strong based cationic resins decreases with increasing acid concentration. Kimura and Ishimori (1958) studied the cation exchange behavior of polonium in nitric acid solutions containing varying amounts of ethylenediamine tetraacetic acid. The distribution coefficient decreases as the amount of reagent increases, which indicates the formation of polonium diethylenediaminetetraacetate.

2.10 Other behavior

Zikovsky (1998) precipitated polonium from solution using barium sulfate as the carrier. Polonium can be isolated from concentrated hydrochloric acid solution using metallic tellurium as the carrier (Lally, 1992). Rushing (1966) also used tellurium as the carrier to coprecipitate polonium from water samples using stannous chloride.

Morrow et al. (1954) studied the colloidal properties of polonium in solution using filtration. In markedly acid or alkaline regions, polonium was completely filterable. In the neutral region (pH 6–10), filterability was low and polonium concentration-dependent, with a marked decrease in filterability occurring as the polonium concentration increased. They believed these observations to be due to alterations in ionic strength.

Bagnall (1957b) described the formation of polonium colloids as being due to the presence of extremely insoluble hydrolysis products resulting from polonium salts originally present in solution. The difficulty with this explanation is that the solubility product of the polonium compound formed by the hydrolysis would never be exceeded in any of the trace level observations recording colloid formation, since the amounts of polonium present are extremely small (Bagnall, 1957a,b). It is more probable that impurities act as centers of adsorption for the material.

Cataphoresis experiments have shown that polonium forms both cations and anions in hydrochloric, nitric, and sulfuric acids. Equal concentrations were observed in $\sim 0.05 \text{ mol L}^{-1}$ solutions of any of the three acids. In acid solutions above 0.5 mol L^{-1}, polonium was present almost completely in the form of ions which migrated toward the anode. In weakly acidic and neutral solutions, polonium exhibited a greater tendency to migrate toward the cathode. This may be due to the formation of charged colloidal particles (Moyer, 1956).

Ulrich and Degueldre (1993) described polonium(IV) as the predominant redox state under oxic conditions ($P_{(O_2)} = 0.2$ bar), existing in strongly acidic media only. The presence of hydroxo-species suggests that polonium may form colloids in neutral or slightly acidic solutions (e.g., $PoO(OH)_2 \cdot PoO_2$), whereas in alkaline media, PoO_3^{2-} predominates.

There is extensive literature on tracer experiments, summarized below, carried out with extremely dilute solutions of polonium, although the inferences drawn are speculative owing to the unavoidable absence of analytical data. The electrodeposition of polonium from hydrochloric acid is principally anodic, even in 0.2 mol L^{-1} solution. Diffusion and ion mobility studies in more dilute acid indicate the presence of divalent cations, the species involved is probably a partially ionized or hydrolyzed form of polonium tetrachloride, $PoCl_2^{2+}$ or PoO^{2+}. Solvent extraction data obtained with dithizone suggest the latter; tracer amounts in nitric acid and milligram amounts in hydrochloric acid probably form $PoODz_2$. A soluble basic chloride is thought to be present in 1–3 mol L^{-1} hydrochloric acid, and hydrolysis to the hydroxide in very dilute solutions has been postulated. However, this is not in agreement with results on the milligram scale which indicate some complex ion formation even in 1 mol L^{-1} acid. Tracer coprecipitation work has shown that very little polonium is carried down on silver chloride at high concentrations of acid or chloride and that it is not coprecipitated with lead or mercurous chloride (Bagnall, 1957a).

Adsorption and desorption experiments were carried out using montmorillonite (a clay). Polonium was sorbed by two different processes, either from solution or generated by the radioactive decay of adsorbed ^{210}Pb. Since the local chemical environment of polonium at the clay surface may be different according to which of the two processes has taken place, the two adsorptive processes could, therefore, lead to a difference in desorption behavior. Polonium adsorption was independent of the ionic strength and thus dominated by strong specific (covalent) interaction with the clay surface. The sorption was quasiirreversible for the species adsorbing from solution, whereas desorption of polonium generated by the decay of ^{210}Pb was significantly enhanced (R_d one order of magnitude less) (Ulrich and Degueldre, 1993).

Polonium is removed from weakly acid (pH 4) solutions of radiolead nitrate by filtration through a bed of titanium dioxide and recovered by washing with 3–6 mol L^{-1} hydrochloric acid. The mechanism involved, however, is uncertain. It is also removed on manganese dioxide from nitric acid. Trace polonium has been carried by bismuth hydroxide in ammoniacal solution; results indicate that an adsorption process is involved. It is also carried on ferric, praseodymium/neodymium, lanthanum, and aluminum hydroxides in alkaline solution (Bagnall, 1957b).

References

Abakumov, A.S., 1982. Thermal reactions of polonium. Russ. Chem. Rev. 51, 622–629.
Ampelogova, N.I., 1973. Ion-exchange study of the complexing of polonium. Radiokhimiya 15, 813–820.
Bagnall, K.W., 1957a. The chemistry of polonium. Quart. Rev. (London) 11, 30–48.
Bagnall, K.W., 1957b. Chemistry of the Rare Radioelements. Butterworths Scientific Publications, London.
Bagnall, K.W., 1983. The chemistry of polonium. Radiochim. Acta 32, 153–161.
Bagnall, K.W., D'Eye, R.W.M., 1954. The preparation of polonium metal and polonium dioxide. J. Chem. Soc. 4295–4299.
Bagnall, K.W., Freeman, J.H., 1956a. Electrochemical studies on polonium. J. Chem. Soc. 2770–2774.
Bagnall, K.W., Freeman, J.H., 1956b. The sulphates and selenate of polonium. J. Chem. Soc. 4579–4582.
Bagnall, K.W., Freeman, J.H., 1957. Solubility of some polonium compounds. J. Chem. Soc. 2161–2163.
Bagnall, K.W., Robertson, D.S., 1957a. Solvent extraction studies with polonium. J. Chem. Soc. 509–512.
Bagnall, K.W., Robertson, D.S., 1957b. Polonium monosulphide. J. Chem. Soc. 1044–1046.
Bagnall, K.W., D'Eye, R.W.M., Freeman, J.H., 1955a. The polonium halides. Part I. Polonium chlorides. J. Chem. Soc. 2320–2326.
Bagnall, K.W., D'Eye, R.W.M., Freeman, J.H., 1955b. The polonium halides. Part II. Bromides. J. Chem. Soc 3959–3963.
Bagnall, K.W., D'Eye, R.W.M., Freeman, J.H., 1956. The polonium halides. Part III. Polonium tetraiodide. J. Chem. Soc. 3385–3389.
Bagnall, K.W., Freeman, J.H., Robertson, D.S., Robinson, P.S., Stewart, M.A.A., 1958a. The Polonium Chemistry Project. UK Atomic Energy Authority, AERE C/R 2566.
Bagnall, K.W., Robertson, D.S., Stewart, M.A.A., 1958b. The polonium nitrates. J. Chem. Soc. 3633–3636.
Beamer, W.H., Maxwell, C.R., 1949. Physical properties of polonium. II. Studies and crystal structure. J. Chem. Phys. 17, 1293–1298.
Borg, R.J., Dienes, G.J., 1981. On the validity of single atom chemistry. J. Inorg. Nucl. Chem. 43, 1129–1133.
Brooks, L.S., 1955. The vapour pressure of polonium. J. Am. Chem. Soc. 77, 3211.
Cairo, A., 1958. Separation of polonium with diisopropylketone. In: Proceedings of the 2nd International Conference on the Peaceful Uses of Atomic Energy, Geneva, 7, pp. 331–335.
Casella, V.R., Bishop, C.T., Glosby, A.A., Hiatt, M.H., Mathews, N.F., Bunce, L.A., et al., 1982. Separation of polonium by ion exchange chromatography. Radiochem. Radioanal. Lett. 55, 279–288.
Charles, G.W., Hunt, D.J., Pish, G., Timma, D.L., 1955. Preliminary description and analysis of the spectrum of polonium. J. Opt. Soc. Am. 45, 869–872.
Charlot, G., 1958. Oxidation-reduction potentials. Pergamon Press, London.

Danon, J., Zamith, A.A.L., 1957. Solvent extraction of polonium from nitric acid. Nature 177, 746–747.

Eichelberger, J.F., Grove, G.R. and Jones, L.V., 1965. Mound Laboratory Progress Report for March 1965. Mound Laboratory Report, MLM-1250.

Figgins, P.E., 1961. The Radiochemistry of Polonium. US Atomic Energy Commission, NAS-NS-3037.

Finkelnburg, W., Stern, F., 1950. Electron screening and ionisation potentials of neutral and singly ionised atoms. Phys. Rev. 77, 303–304.

Fry, C., Thoennessen, M., 2013. Discovery of actinium, thorium, protactinium, and uranium isotopes. Atom. Data Nucl. Data Tables 99, 345–364.

Goode, J.M., 1956. Physical properties of polonium. In: Moyer, H.V., Gnagey, L.B., Rogers, A.J. (Eds.), Polonium. US Atomic Energy Commission, pp. 18–32. , TID-5221.

Greenwood, N.N., Earnshaw, A., 1998. Chemistry of the Elements. Butterworth-Heinemann, Oxford.

Hataye, I., Suganuma, H., Sakata, M., Nagame, Y., 1981. Solvent extraction study on the hydrolysis of tracer concentration of polonium(IV) in perchlorate solutions. J. Inorg. Nucl. Chem. 43, 2101–2104.

Ishimori, T., 1955. Separation of RaD, RaE and RaF by ion exchange. Bull. Chem. Soc. Japan 28, 432–435.

Joliot, F., 1930. Étude électrochimque des radioéléments applications diverses. J. Chim. Phys. 27, 9–159.

Kendall, J., Andrews, J.C., 1921. The solubilities of acids in aqueous solutions of other acids. J. Amer. Chem. Soc. 43, 1545–1560.

Khlopin, V.G., Samartseva, A.G., 1934. Studies on the chemistry of polonium. I. Some compounds of bivalent polonium. Dokl. Akad. Nauk SSSR 4, 433–436.

Kimura, K., Ishimori, T., 1958. Some studies on the tracer chemistry of polonium. In: Proceedings of the 2nd International Conference on the Peaceful Uses of Atomic Energy, Geneva, 28, pp. 155–155.

Kocher, D.C., 1977. Nuclear Decay Data for Radionuclides Occurring in Routine Releases from Nuclear Fuel Cycle Facilities. Oak Ridge National Laboratory, ORNL/NUREG/T102.

Lally, A.E., 1992. Chemical procedures. In: Ivanovich, M., Harmon, R.S. (Eds.), Uranium-Series Disequilibrium: Applications to Earth, Marine and Environmental Sciences, second ed. Clarendon Press, Oxford, pp. 95–126.

Latimer, W.M., 1952. The Oxidation States of the Elements and Their Potentials in Aqueous Solutions., second ed. Prentice-Hall, Englewood Cliffs, NJ.

Lo, J.M., Wai, C.M., 1983. Determination of extraction constant for polonium diethyldithiocarbamate by substoichiometric extraction. Anal. Chim. Acta 148, 327–330.

Marechal-Cornil, J., Picciotto, E., 1953. Séparation des radioelements naturels par l'oxyde de mésityle. Bull. Soc. Chim. Belg. 62, 372–382.

Marsh, S.F., Alarid, J.E., Hammond, C.F., McLeod, M.J., Roensch, F.R., Rein, J.E., 1978a. Cation exchange of 53 elements in nitric acid. Los Alamos Scientific Laboratory Report, LA-7083.

Marsh, S.F., Alarid, J.E., Hammond, C.F., McLeod, M.J., Roensch, F.R., Rein, J.E., 1978b. Anion exchange of 58 elements in hydrobromic acid and in hydroiodic acid. Los Alamos Scientific Laboratory Report, LA-7084.

Matsuura, N., Haissinsky, M., 1958. Sur la valence six du polonium. J. Chim. Phys. 55, 475–482.

Maxwell, C.R., 1949. Physical properties of polonium. I. Electrical resistance, density and allotropy. J. Chem. Phys. 17, 1288–1292.

Moore, F.L., 1957. Long-chain amines: versatile acid extractants. Anal. Chem. 29, 1660–1662.

Morrow, P.E., Della Rosa, R.J., Casarett, L.J., Miller, G.J., 1954. A Preliminary Investigation of Some Radiocolloidal Properties of Polonium-210 Using Molecular Filters. US Atomic Energy Commission, UR-363.

Moyer, H.V., 1956. Chemical properties of polonium. In: Moyer, H.V., Gnagey, L.B., Rogers, A.J. (Eds.), Polonium. US Atomic Energy Commission, pp. 33–96. , TID-5221.

Nelson, F., Murase, T., Kraus, K.A., 1964. Ion exchange procedures: 1. Cation exchange in concentrated HCl and $HClO_4$ solutions. J. Chromatog. 13, 503–535.

Nikol'skii, B.P., Sinitsyna, G.S., Ziv, D.M., 1958. The determination of the valency of polonium in solution. Trudy Rad. Inst. Im. V.G. Khlop 8, 141–152.

Power, W.H., 1949a. Quarterly progress report. Mound Laboratory Report 62. MLM-368-1.

Power, W.H., 1949b. Quarterly progress report. Mound Laboratory Report 82. MLM-379-1.

Reischmann, F.J., Trautmann, N., Herrmann, G., 1984. Chemistry at low concentrations: polonium at a level of 5000 to 40 atoms. Radiochim. Acta 36, 139–143.

Reischmann, F.J., Rumler, B., Trautmann, N., Herrmann, G., 1986. Chemistry at low concentrations: polonium at a level of 5000 to 40 atoms. Radiochim. Acta 39, 185–188.

Rudolph, J., Bächmann, K., 1979. Determination of adsorption enthalpies and entropies of inorganic halides by temperature-programmed gas chromatography. J. Chromatogr. 178, 459–469.

Rushing, D.E., 1966. Determination of dissolved polonium-210 in water by coprecipitation with tellurium by stannous chloride. Anal. Chem. 38, 900–905.

Sasaki, Y., 1955. Separation of sixth B group elements (S, Se, Te, Po) with anion exchange resin. Bull. Chem. Soc. Japan 28, 89.

Schulz, W.W., Navratil, J.D., Bess, T. (Eds.), 1987. Science and Technology of Tributyl Phosphate. Volume II – Selected Technical and Industrial Uses. Part B. CRC Press, Boca Raton, FL.

Shannon, R.D., 1976. Revised effective ionic radii and systematic studies of interatomic distances in halides and chalcogenides. Acta Cryst. A32, 751–767.

Sillén, L.G., Martell, A.E., 1964. Stability Constants of Metal–Ion Complexes. The Chemical Society, London, Special Publication No. 17.

Starik, I.E., Ampelogova, N.I., Kuznetsov, B.S., 1964a. Hydrolysis of polonium in perchloric acid solutions. Radiokhimiya 6, 519–524.

Staritzky, E., 1951. Some Compounds of Tetravalent Polonium. US Atomic Energy Commission, LA-1286.

Stull, D.R., Sinke, G.C., 1956. Thermodynamic Properties of the Elements., vol. 18. American Chemical Society.

Suganuma, H., 1995. Anion-exchange of the chemical species of tracer concentrations of polonium(IV) in chloride solutions. J. Inorg. Nucl. Chem., Articles 191, 265–272.

Suganuma, H., Hataye, I., 1981. Solvent extraction study on the hydrolysis of tracer concentration of Po(IV) in chloride solutions. J. Inorg. Nucl. Chem. 43, 2511–2515.

Tompkins, E.R., 1951. Ion Exchange Experiments With Polonium. US Atomic Energy Commission, UCRL-1294.

Ulrich, H.J., Degueldre, C., 1993. The sorption of ^{210}Pb, ^{210}Bi and ^{210}Po on montmorillonite: a study with emphasis on reversibility aspects and on the effect of the radioactive decay of adsorbed nuclides. Radiochim. Acta 62, 81–90.

Van Muylder, J., 1966. Polonium. In: Pourbaix, M. (Ed.), Atlas of Electrochemical Equilibria in Aqueous Solutions. Pergamon Press, Oxford, pp. 572–576. , Sect. 19.5.

Wahl, A.C., Bonner, N.A. (Eds.), 1951. Radioactivity Applied to Chemistry. John Wiley & Sons, New York.

Wai, C.M., Lo, J.M., 1982. Extraction and separation of ^{210}Pb, ^{210}Bi and ^{210}Po by diethyldithiocarbamate. Radiochem. Radioanal. Lett. 50, 293–298.
Weinstock, B., Chernick, C.L., 1960. Preparation of a volatile polonium fluoride. J. Am. Chem. Soc. 82, 4116–4117.
Winter, M., 2000. http://www.webelements.com, December 12, 2000 update.
Witteman, W.G., Giorgi, A.L., Vier, D.T., 1960. The preparation and identification of some intermetallic compounds of polonium. J. Phys. Chem. 64, 434–440.
Zhdanov, S.I., 1985. Sulfur, selenium, tellurium, and polonium. In: Bard, A.J., Parsons, R., Jordan, J. (Eds.), Standard Potentials in Aqueous Solution. Marcel Dekker Inc, New York, pp. 93–125.
Zikovsky, L., 1998. Precipitation and solubility of some polonium compounds. J. Radioanal. Nucl. Chem. 227, 171–172.

CHAPTER 3

Chemical thermodynamics of polonium

3.1 General principles

The aqueous chemistry of polonium is poorly understood. To be able to predict the behavior of polonium in solution, and hence develop methods for its effective removal or control, it is necessary to develop an understanding, or indication, of the aqueous speciation of the element. For example, is polonium present as a chloride or sulfate complex, are the complexes cationic or anionic, or a mixture? Also, depending on conditions, polonium may be present as a colloid. This understanding can be achieved through a combination of experimental techniques and the development of a thermochemical database for polonium, which can be used to predict speciation characteristics of the element in solution.

The speciation of an element in solution is governed by the elemental composition of the aqueous solution, that is, the list of possible chemical species which may form in the solution is based on the chemical elements in solution. For example, the following species set ((aq) omitted for charged species), Ca^{2+}, CO_3^{2-}, H_2O, H^+, OH^-, HCO_3^-, $CaHCO_3^+$, CO_2(g), H_2CO_3(aq), $CaCO_3$(aq), $CaCO_3$(s), $CaOH^+$, and $Ca(OH)_2$(s), may be possible when a solution contains calcium and carbonate. Water (H_2O) and its associated species, H^+ and OH^-, are always present since water is the solvating medium. The formation of each of the individual species is dependent on their relative stability (as controlled by their free energy of formation), the concentrations of the reacting elements/ions and the solution characteristics (e.g., temperature, redox potential). To understand the aqueous chemistry of any element, a list of species that are likely to form, given a particular set of reacting elements and/or ions, is required.

The Gibbs energy of formation ($\Delta G_f°$) values for polonium species have been derived using the thermochemical equations described in the following sections in this chapter. The following values have been used in calculations involving the Nernst equation.

The Aqueous Chemistry of Polonium and the Practical Application of its Thermochemistry
DOI: https://doi.org/10.1016/B978-0-12-819308-2.00003-6

$$R = 8.314\,\mathrm{J\,mol^{-1}\,K^{-1}}$$

$$T = 298.15\mathrm{K}$$

$$F = 96,487\ \mathrm{C\,mol^{-1}}$$

For brevity, only the chemical formulae of other reacting species used in the calculations are contained in the text. All of the relevant thermochemical data for these species can be found in Appendix 1. Similarly, the thermochemical data for polonium species will only be given where those particular data are derived.

Electrochemical reactions in aqueous solution can be represented graphically by pH-potential diagrams as predicted on the basis of thermodynamic equilibria. For every chemical reaction, there is a corresponding "free energy of reaction," ΔG_r, whose value and sign determine the tendency of the system to react and the direction of the reaction. Under fixed physical conditions, ΔG_r can be expressed as the sum of a standard free energy (which is constant for fixed physical conditions characteristic of a standard reference state) and a linear combination of the logarithms of the activities of those substances taking part in the reaction (Piontelli, 1966). All values quoted throughout this section refer to standard conditions of $P = 101$ kPa and unit activities, unless otherwise stated.

For systems whose constituent species are a particular element in aqueous solution, water, and its constituents (H^+, OH^-, gaseous hydrogen, and oxygen) and the products of the element with these species (oxides, hydroxides, hydrides, etc.) either as a solid or in solution, the equilibrium conditions of the electrochemical reactions can be represented by a relationship (Eq. (3.1)) between the equilibrium potential, the standard free energy of reaction (the standard equilibrium potential, $E°$) and a linear combination of the logarithms of the activities using the Nernst equation,

$$E = E^{\circ} + \frac{2.303RT}{nF}\log\frac{[a_{\mathrm{ox}}]}{[a_{\mathrm{red}}]} \qquad (3.1)$$

where E is the potential (V), $E°$ the standard electrode potential (V), R the universal gas constant (8.314 J mol^{-1} K^{-1}), T the temperature (K), n the number of electrons involved in the reaction, F is the Faraday constant (96,487 C mol^{-1}), $[a_{\mathrm{ox}}]$ the activity of the oxidized form and $[a_{\mathrm{red}}]$ the activity of the reduced form.

With respect to Eq. (3.1), the term

$$\log\frac{[a_{ox}]}{[a_{red}]}$$

corresponds to $\log {K_T}^{-1}$ (where K_T is the equilibrium constant) for reactions written as reductions.

In the case of nonelectrochemical reactions, the equilibrium conditions assume a simpler form as no equilibrium potential is involved. These conditions are then expressed by a relationship between the standard free energy of reaction (the equilibrium constant, $K_T°$; written here as $K°$) and a linear combination of the logarithms of the activities (Piontelli, 1966) viz,

$$\log K° = \Sigma \log a_{products} - \Sigma \log a_{reactants} \tag{3.2}$$

where $\Sigma \log a_{products}$ and $\Sigma \log a_{reactants}$ are the sum of the logarithms of the activities of the products and reactants, respectively.

The ΔG_r value for a given chemical reaction can be calculated from either the $E°$ or $K°$ value using equations Eqs. (3.3) and (3.4) and is related to the Gibbs energies of formation, ΔG_f, via Eq. (3.5).

$$\Delta G_r° = -nFE° \tag{3.3}$$

$$\Delta G_r° = -2.303RT \log K° \tag{3.4}$$

$$\Delta G_r° = \Sigma \Delta G_f°(\text{products}) - \Sigma \Delta G_f°(\text{reactants}) \tag{3.5}$$

The pH-potential diagram produced for a given set of electrochemical reactions allows the predominance of individual chemical species to be determined at specific conditions of pH and redox potential.

As indicated, values of $E\underline{o}$ may be obtained from nonelectrochemical measurements of the standard Gibbs energy change of a reaction according to Eq. (3.3). Similarly, they may also be obtained from the equilibrium constant of a reaction by a combination of Eqs. (3.3) and (3.4). In this way, values may be obtained for electrodes that are difficult or impossible to set up experimentally.

Standard Gibbs energies may be obtained by purely thermal methods using the relationship described by Eq. (3.6),

$$\Delta G\underline{o} = \Delta H\underline{o} - T\Delta S\underline{o} \tag{3.6}$$

where $\Delta G⦵$ is the Gibbs energy change of reaction (kJ mol^{-1}), $\Delta H⦵$ the enthalpy of reaction (kJ mol^{-1}), T the temperature (K) and $\Delta S⦵$ the entropy of reaction (J mol^{-1} K^{-1}).

In the case of ions in solution, it is necessary to measure an equilibrium reaction between the ions and a substance whose entropy is known. This is most often a solubility equilibrium, where the equilibrium between an ionic crystal and its ions are studied in solutions of various concentrations of a salt whose ions do not take part in the equilibrium. By extrapolation to zero ionic strength, the thermodynamic constant can be found and the standard Gibbs energy of the reaction calculated using Eq. (3.4).

As described by Parsons (1985), all experimental measurements relate to reactions in which electrical charge is conserved, and therefore it is not possible to obtain the thermodynamic properties of single ionic species. For reactions such as those described by Eqs. (3.7) and (3.8), the standard thermodynamic properties conventionally quoted for aqueous solutions are those in which the ion is produced from the element (or elements) in its standard state, while the hydrogen ion is either removed from, or added to, solution and transformed to, or from, molecular hydrogen under standard conditions, such that the total charge of the system remains unchanged.

$$\frac{1}{2}Cl_2(g) + \frac{1}{2}H_2(g) \rightleftharpoons Cl^-(aq) + H^+(aq) \tag{3.7}$$

$$Po(s) + 2H^+(aq) \rightleftharpoons Po^{2+}(aq) + H_2(g) \tag{3.8}$$

Convention states that the enthalpy and Gibbs energy of an element in its standard state are taken to be zero, and so, it follows that the standard enthalpy change ($\Delta H_f°$) in reactions (3.7) and (3.8) are represented by Eqs. (3.9) and (3.10), respectively.

$$\Delta H_f° = H_f°(Cl^-(aq)) + H_f°\left(H^+(aq)\right) \tag{3.9}$$

$$\Delta H_f° = H_f°\left(Po^{2+}(aq)\right) - 2H_f°\left(H^+(aq)\right) \tag{3.10}$$

For clarity, the (aq) subscript is omitted for hydrated ions, and therefore the general equation for the standard enthalpy of an ionic species, B, having an ionic charge of z_B at 298.15K is given in Eq. (3.11).

$$\Delta H_f°(B) = H_f°(B) - z_B H_f°(H^+) \quad (3.11)$$

Similarly, the standard Gibbs energy, $\Delta G_f°$, of B at 298.15K is,

$$\Delta G_f°(B) = G_f°(B) - z_B G_f°(H^+) \quad (3.12)$$

In view of the convention for the standard entropy (i.e., that the entropy is zero for a perfectly ordered pure solid at 0K), the corresponding standard ionic entropy must be calculated from the entropy change in reactions like (3.7) and (3.8), taking into account the entropies of the elements in their standard states. For example, at 298.15K,

$$S°(Cl^-) = S°'(Cl^-) + S°'(H^+) \quad (3.13)$$

$$S°(Po^{2+}) = S°'(Po^{2+}) - 2S°'(H^+) \quad (3.14)$$

or

$$S°(Cl^-) = S° + \frac{1}{2} S°(H_2) + \frac{1}{2} S°(Cl_2) \quad (3.15)$$

$$S°(Po^{2+}) = S° - S°(H_2) + S°(Po) \quad (3.16)$$

where the prime indicates the "absolute" entropy of the hydrated ions and ΔS_{O} is the respective standard entropy change in reactions (3.7) and (3.8). The general definition of the standard entropy of B at 298.15K is, therefore

$$S°(B) = S°'(B) - z_B S°'(H^+) \quad (3.17)$$

Owing to the conventional definition of the standard enthalpy of elements, these quantities also relate the change in enthalpy (energy) accompanying a real process like (3.7) or (3.8). However, in the case of the entropy, a different reference state for the element is used and the entropy change in these real processes at 298.15K must be calculated from Eq. (3.18),

$$\Delta S° = S°(B) + \left(\frac{z_B}{2}\right) S°(H_2) - \left(\frac{1}{n}\right) S_e°(B_n) \quad (3.18)$$

where $S_e°(B_n)$ is the entropy of element B at 298.15K and n is the stoichiometric coefficient of the element in its standard state.

For compounds or complexes, the expression for the entropy of reaction is extended to take into account each element comprising the compound or complex. For example, in the formation of $PoCl_4^{2-}$, as given by Eq. (3.19),

$$Po(s) + 2Cl_2(g) + H_2(g) \rightleftharpoons PoCl_4^{2-}(aq) + 2H^+(aq) \tag{3.19}$$

the entropy of reaction is given by Eqs. (3.20) and (3.21).

$$S^\circ\left(PoCl_4^{2-}\right) = S^{\circ\prime}\left(PoCl_4^{2-}\right) + 2S^{\circ\prime}\left(H^+\right) \tag{3.20}$$

$$S^\circ\left(PoCl_4^{2-}\right) = \Delta S^\circ + S^\circ(Po) + 2S^\circ(Cl_2) + S^\circ(H_2) \tag{3.21}$$

3.1.1 Ionic strength corrections

The standard state for a solute B in a solution at standard pressure exists when the molal concentration, m_B, is equal to m° or 1 mol kg^{-1} and the activity coefficient, γ_B, is unity (Grenthe et al., 1992). For many reactions, however, measurements cannot be made accurately or at all in dilute solutions, from which extrapolation to the standard state would be simple. Precise thermodynamic information for these systems can only be obtained in the presence of an inert electrolyte of sufficiently high concentration, to ensure that activity factors are reasonably constant throughout the measurements.

Thermodynamic data at the standard state (ionic strength, I_m, equals zero) is obtained from extrapolation of data acquired in high ionic strength media by estimating the activity coefficients of all species participating in the relevant reaction. The Debye-Hückel term, D, that is the dominant term in the expression for the activity coefficients in dilute solution, accounts for electrostatic, nonspecific long-range interactions. At higher concentrations, short-range, nonelectrostatic interactions have to be considered, and this is done by adding ionic strength dependent terms to the Debye-Hückel expression (Eq. (3.22)):

$$D = \frac{A\sqrt{I_m}}{1 + Ba_j\sqrt{I_m}} \tag{3.22}$$

where A and B are constants that are temperature dependent, and a_j is the effective diameter of the hydrated ion, j. At 25°C, the values of A and Ba_j are 0.509 and 1.5 kg$^{1/2}$ mol$^{-1/2}$, respectively (Grenthe et al., 1992).

The specific ion interaction theory has been used to relate data obtained in high ionic strength media to the standard state. It is based on the following two assumptions:

1. The activity coefficient γ_j of an ion j of charge z_j in the solution of ionic strength I_m may be described by Eq. (3.23):

$$\log\gamma_j = -z_j^2 D + \sum_k \varepsilon_{(j,k,I_m)} m_k \quad (3.23)$$

where m_k is the molality of ion k of the ionic medium.

2. The ion interaction coefficients $\varepsilon_{(j,k,I_m)}$ are zero for ions of the same charge sign and for uncharged species.

When using the specific ion interaction theory, the relationship between the redox potential of a couple in a medium of ionic strength and the corresponding quantity at $I_m = 0$, can be calculated in the following way. For reaction (3.24),

$$PoCl_4^{2-} + H_2 \rightleftharpoons Po(s) + 2H^+ + 4Cl^- \quad (3.24)$$

the thermodynamic expression for the stability constant is given by Eq. (3.25):

$$K^o = \frac{a_{Po} a_{H^+}^2 a_{Cl^-}^4}{a_{PoCl_4^{2-}} a_{H_2}} \quad (3.25)$$

where a_j is the activity of species j in the aqueous solution. The activity of j can be related to its concentration in solution by:

$$a_j = [j] \cdot \gamma_j \quad (3.26)$$

Thus Eq. (3.25) can be expressed as:

$$K^o = \frac{[Po][H^+]^2[Cl^-]^4}{[PoCl_4^{2-}][H_2]} \frac{\gamma_{Po}\gamma_{H^+}^2\gamma_{Cl^-}^4}{\gamma_{PoCl_4^{2-}}\gamma_{H_2}} \quad (3.27)$$

Eq. (3.27) can be rewritten as:

$$K^o = K\frac{\gamma_{Po}\gamma_{H^+}^2\gamma_{Cl^-}^4}{\gamma_{PoCl_4^{2-}}\gamma_{H_2}} \quad (3.28)$$

where K is the stability constant measured at the ionic strength, I_m. Taking logarithms gives:

$$\log K^{\circ} = \log K + \log\gamma_{Po} + 2\log\gamma_{H^+} + 4\log\gamma_{Cl^-} - \log\gamma_{PoCl_4^{2-}} - \log\gamma_{H_2} \tag{3.29}$$

The activity of solid Po is unity and at low partial pressures of H_2, $\log\gamma_{H_2} = 0$ (Grenthe et al., 1992). Therefore Eq. (3.29) simplifies to:

$$\log K^{\circ} = \log K + 2\log\gamma_{H^+} + 4\log\gamma_{Cl^-} - \log\gamma_{PoCl_4^{2-}} \tag{3.30}$$

In a medium of HCl, for example, the specific ion interaction theory defines the activity coefficients for the species listed in Eq. (3.30), as follows:

$$\log\gamma_{H^+} = -D + \varepsilon_{(H^+,Cl^-)} m_{Cl^-} \tag{3.31}$$

$$\log\gamma_{Cl^-} = -D + \varepsilon_{(H^+,Cl^-)} m_{H^+} \tag{3.32}$$

$$\log\gamma_{PoCl_4^{2-}} = -4D + \varepsilon_{(H^+,PoCl_4^{2-})} m_{H^+} \tag{3.33}$$

Therefore Eq. (3.30) can be rewritten as:

$$\log K^{\circ} = \log K - 2D + \Delta\varepsilon\, m_{Cl^-} \quad (\text{since } m_{Cl^-} = m_{H^+}) \tag{3.34}$$

where

$$\Delta\varepsilon = 6\varepsilon_{(H^+,Cl^-)} - \varepsilon_{(H^+,PoCl_4^{2-})} \tag{3.35}$$

Rearranging Eq. (3.34) gives:

$$\log K - 2D = \log K^{\circ} - \Delta\varepsilon\, m_{Cl^-} \tag{3.36}$$

By definition, at equilibrium, log K° can also be equated to $nFE^{\circ}/2.303RT$, and therefore Eq. (3.36) can be rewritten as:

$$E - 2\left(\frac{\ln(10)RT}{nF}\right)D = E^{\circ} - \Delta\varepsilon\left(\frac{\ln(10)RT}{nF}\right)m_{Cl^-} \tag{3.37}$$

For reaction (3.24) (i.e., $n = 2$), and at 25°C,

$$E - 59.16\,D = E^{\circ} - 29.58\Delta\varepsilon\, m_{Cl^-} \tag{3.38}$$

By plotting the molality of chloride (m_{Cl^-}) against the corrected oxidation potential, $E - 59.16D$, a straight line with an intercept of E° and a slope of $-29.58\,\Delta\varepsilon$ results.

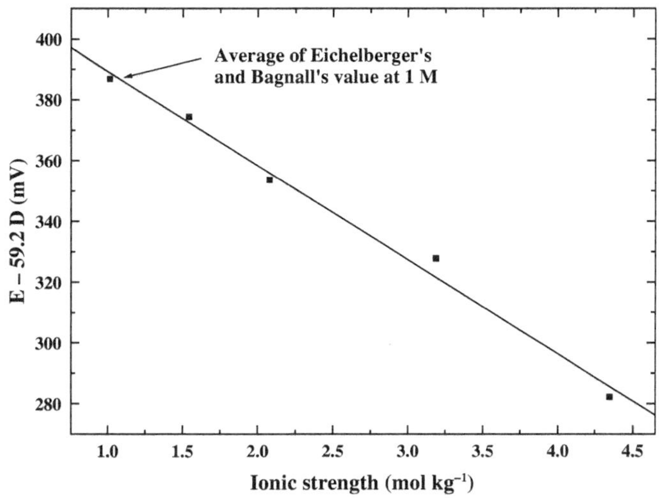

Figure 3.1 Plot of ionic strength versus E - 59.2D for the $PoCl_4^{2-}$/Po oxidation couple.

Table 3.1 Data used to derive $E°$ and $\Delta\varepsilon$.

Molarity (c_B/mol L^{-1})	Conversion factor	Molality (m_B/mol kg^{-1})	E (mV)	D (mV)	E − 59.2D (mV)	Reference
1	1.0222	1.0222	417	0.204	405	Eichelberger et al. (1965)
1.5	1.0324	1.5486	387	0.221	374	Eichelberger et al. (1965)
2	1.0430	2.0860	367	0.232	353	Eichelberger et al. (1965)
3	1.0654	3.1962	342	0.247	327	Eichelberger et al. (1965)
4	1.0893	4.3572	297	0.257	282	Eichelberger et al. (1965)
1	1.0222	1.0222	380	0.204	368	Bagnall and Freeman (1956a)

This is illustrated in Fig. 3.1 for the $PoCl_4^{2-}$/Po oxidation couple using the concentration data summarized in Table 3.1, obtained from Eichelberger et al. (1965) and Bagnall and Freeman (1956a). The factors for the conversion of molarity, c_B, to molality, m_B, were obtained from Grenthe et al. (1992). The line of best fit shows good agreement between the parameters and has an intercept, $E°$, of 0.420 ± 0.005 V and a slope, $-29.58\ \Delta\varepsilon$, of -31 ± 2 [$r^2 = 0.9909$; $P \leq 0.001$; $n = 5$].

3.2 Thermochemical properties for polonium species

3.2.1 Po(s), Po(g), and Po_2(g)

Polonium metal is the standard state for polonium and, by definition, both $\Delta G_f°$ and heat of formation, $\Delta H_f°$, are:

$$\Delta G_f°(\text{Po(s)}) = 0.0\ \text{kJ mol}^{-1}$$

$$\Delta H_f°(\text{Po(s)}) = 0.0\ \text{kJ mol}^{-1}$$

Stull and Sinke (1956) estimated a value for the entropy of formation, $S_f°$, of polonium metal:

$$S_f°(\text{Po(s)}) = (62.8 \pm 2.1)\ \text{J mol}^{-1}\ \text{K}^{-1}$$

The uncertainty was assigned by Kelley and King (1961). Zhdanov (1985) selected the entropy value of Stull and Sinke in a critical review of the thermodynamics of polonium.

Brooks (1955) measured the vapor pressure of polonium gas above metallic polonium from 711K to 1018K. The vapor pressure measurements were described with Eq. (3.39):

$$\log P = 7.235 - \frac{5378}{T} \tag{3.39}$$

where P is the pressure in mm (Hg) and T is the temperature in kelvin. The behavior is consistent with that found by Abakumov and Ershova (1974) over a similar temperature range. Stull and Sinke (1956), however, indicated that the data of Brooks were best interpreted by assuming both a monoatomic (Po(g)) and diatomic (Po_2(g)) species in the gaseous phase. This interpretation is supported by the fact that a diatomic species is important in the gaseous phase of both bismuth and tellurium, the neighboring elements of polonium (Stull and Sinke, 1956). Stull and Sinke estimated thermodynamic values of:

$$\Delta G_f°(\text{Po(g)}) = (106.6 \pm 0.2)\ \text{kJ mol}^{-1}$$

$$\Delta H_f°(\text{Po(g)}) = (144.1 \pm 0.2)\ \text{kJ mol}^{-1}$$

$$\Delta G_f°(Po_2(g)) = (93.0 \pm 0.4)\ kJ\ mol^{-1}$$

$$\Delta H_f°(Po_2(g)) = (137.7 \pm 0.4)\ kJ\ mol^{-1}$$

The uncertainties have been estimated in the present review. Values for $S_f°$ of:

$$S_f°(Po(g)) = (188.80 \pm 0.04)\ J\ mol^{-1}\ K^{-1}$$

$$S_f°(Po_2(s)) = (275.3 \pm 0.1)\ J\ mol^{-1}\ K^{-1}$$

have also been reported by Stull and Sinke (1956) for the two species. For Po(g), Kelley and King (1961) assigned the uncertainty, while the uncertainty for $Po_2(g)$ has been assigned in the present review.

3.2.2 Po^{2-}

Latimer (1952) estimated an oxidation potential for the H_2Po/Po and Po^{2-}/Po couples of $E° < -1.0$ and < -1.4, respectively. Latimer used sulfur data to determine data for selenium and tellurium, and then subsequently, the three sets of data to determine values for polonium.

Kapustinskii (1948) demonstrated that a linear relationship often exists between the $\Delta H_f°$ of a particular type of species in a single group of the periodic table and the logarithm of the atomic number (log A) comprising the species. It can be shown that a similar relationship would hold between $\Delta G_f°$ and log A. Since $E°$ is directly proportional to $\Delta G_f°$ for the X^{2-}/X oxidation couple, where X is S, Se, Te or Po, a plot of $E°$ versus log A may also show a correlation.

Brown (2001) used the oxidation potential values given by Latimer (1952) for S, Se, and Te (see Table 3.2) and the relationship between $E°$ and log A to derive oxidation potentials for both the $H_2Po(aq)/Po(s)$ and $Po^{2-}/Po(s)$ couples. Fig. 3.2 illustrates these relationships and demonstrates that there is an excellent correlation. The intercepts and slopes of the lines of best fit are 2.16 (0.03) and −1.68 (0.02), respectively, for the H_2X/X couple [$r^2 = 0.9999$; $P \leq 0.01$; $n = 3$] and 1.08 (0.06) and −1.30 (0.04), respectively, for the X^{2-}/X couple [$r^2 = 0.9990$; $P \leq 0.05$; $n = 3$]. The oxidation potentials that were determined for Po from these linear relationships were −1.07 and −1.42 for the $H_2Po(aq)/Po(s)$ and $Po^{2-}/Po(s)$ couples, respectively (Brown, 2001). These values were found to be in very good agreement with those proposed by Latimer (1952).

Table 3.2 Oxidation potentials for the H_2X/X and X^{2-}/X couples of O, S, Se, and Te at 25°C and zero ionic strength and respective ionic radii.

Chalcogen	Oxidation potential (V)				Ionic radius (Å) Shannon (1976)
	H_2X/X		X^{2-}/X		
	Latimer (1952)	More recent data[a]	Latimer (1952)	More recent data[a]	
O		1.229			1.40
S	0.14		− 0.48		1.84
Se	− 0.40	− 0.111	− 0.92	− 0.666	1.98
Te	− 0.72	− 0.51	− 1.14	− 0.95	2.21

[a]O data from Sillén and Martell (1964), Se data from Olin et al. (2005), and Te data from Panson (1963).

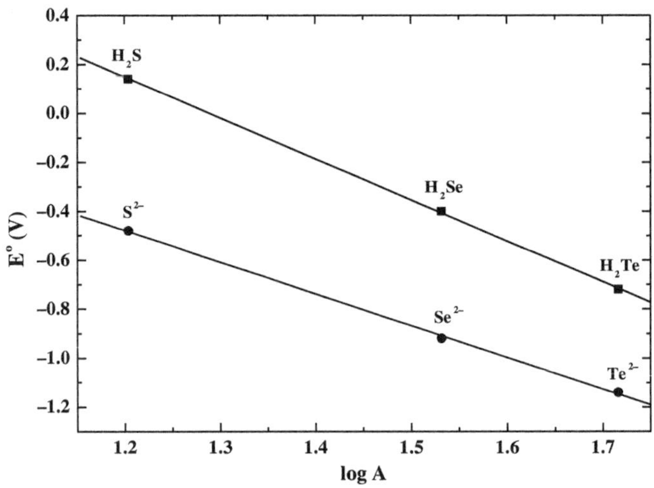

Figure 3.2 Plot of log A versus $E°$ for H_2X/X and X^{2-}/X oxidation couples (X is S, Se, or Te).

Although excellent correlations are obtained from Fig. 3.2, more recent data for Se (Olin et al., 2005) and Te (Panson, 1963) invalidate these relationships and also the methodology of Kapustinskii (1948). A relationship, however, can be developed between the $E°$ values of the oxidation couples and $1/r$, the reciprocal of the ionic radius of the various chalcogenides. Table 3.2 summarizes the oxidation potential data of O, S, Se, and Te from Latimer (1952), Sillén and Martell (1964), Olin et al. (2005), and Panson (1963) and the respective ionic radii, including that of oxygen, from Shannon (1976) that were used to derive Figs. 3.2 and 3.3.

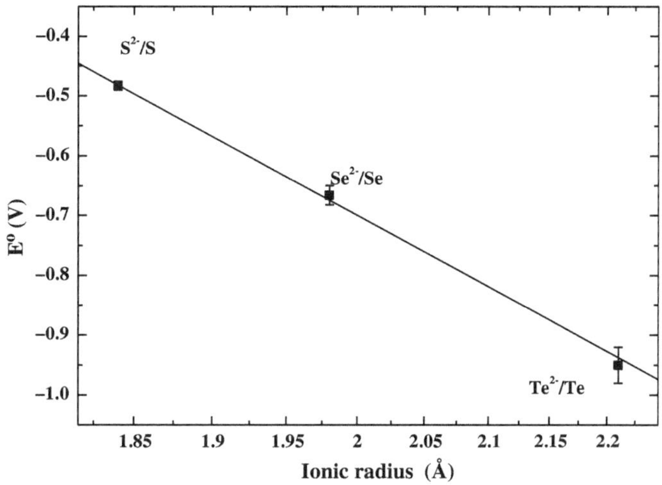

Figure 3.3 Plot of $1/r$ versus $E°$ for the X^{2-}/X oxidation couple (X is S, Se, or Te).

Fig. 3.3 contains plots of $1/r$ of the respective chalcogen versus $E°$ for the X^{2-}/X oxidation couple of S, Se, and Te using the more recent data listed in Table 3.2 for Se and Te. The intercept and slope of the line of best fit are -3.29 ± 0.08 and 5.18 ± 0.16, respectively, for the X^{2-}/X couple [$r^2 = 0.9991$; $P \leq 0.05$; $n = 3$]. The oxidation potential for polonium can be determined from this linear relationship, using the value of 2.284 Å given by Bagnall (1983) for Po^{2-}, and is found to be -1.02 V for the $Po^{2-}/Po(s)$ couple. This value is substantially more positive than the value estimated by Latimer (1952). This trend is also seen in the more recent values for the respective Se and Te oxidation couples, as is shown in Table 3.2.

The applicability of the relationship used in the current work is also demonstrated for the halogen oxidation couples (X^{-}/X) in Fig. 3.4 [the oxidation couples have been taken from Kuhn and Rice (1985) and the ionic radii for the halogens from Shannon (1976)]. The intercept and slope of the line of best fit are -2.92 ± 0.13 and 7.71 ± 0.23, respectively [$r^2 = 0.9982$; $P \leq 0.001$; $n = 4$], indicating an excellent correlation.

From the estimated oxidation potential of -1.02 V for the Po^{2-}/Po (s) couple, the calculated Gibbs energy value is:

$$\Delta G_f°\left(Po^{2-}\right) = (197.1 \pm 7.7)\ \mathrm{kJ\ mol^{-1}}$$

where the uncertainty has been estimated from an uncertainty of 0.04 assigned to the $E°$ of the $Po^{2-}/Po(s)$ couple derived in the present study.

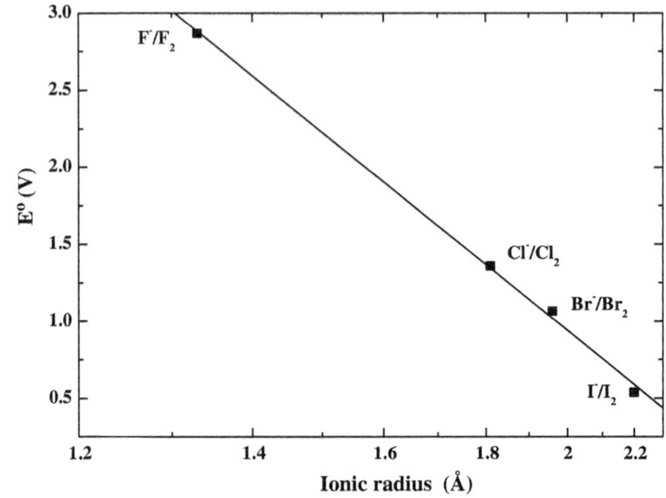

Figure 3.4 Plot of $1/r$ versus $E°$ for the X^-/X oxidation couple (X is F, Cl, Br, or I).

3.2.3 Po^{2+}

Paneth and von Hevesy (1913) and Haissinsky (1932) studied the potentials of the cathodic reduction of polonium to the metal and the anodic oxidation to the dioxide. From their studies, Latimer (1952) derived an E ° value of 0.65 V for the $Po^{2+}/Po(s)$ couple, whilst a value of 0.6 V was obtained by Charlot (1958) from the earlier work of Schneidt (1929). Nikol'skii et al. (1958) also studied the reduction of Po^{2+} to the metal at 18°C and obtained an $E°$ value of 0.68 V for the $Po^{2+}/Po(s)$ couple. For Eq. (3.40),

$$Po^{2+} + 2e^- \rightleftharpoons Po(s) \quad (3.40)$$

an average $E°$ value of 0.643 ± 0.037 V is derived where the uncertainty spans the range in measured values. From this value and the Gibbs energy of Po(s), the Gibbs energy for the polonium(II) ion is:

$$\Delta G_f°(Po^{2+}) = (124.1 \pm 8.4)\ \text{kJ mol}^{-1}$$

3.2.4 $PoO_2(s)$

The $E°$ value for Eq. (3.41) has been quoted as 0.74 (Latimer, 1952), 0.724 (Van Muylder, 1966), and 0.73 V (Zhdanov, 1985).

$$PoO_2(s) + 4H^+ + 4e^- \rightleftharpoons Po(s) + 2H_2O \tag{3.41}$$

Latimer derived his value from the earlier work of Paneth and von Hevesy (1913) and Haissinsky (1932). An average $E°$ value of 0.731 ± 0.009 V is derived, where the uncertainty has been assigned to span the range in values. Zhdanov (1985) also gives values for the $\Delta H_f°$ and $S_f°$ of PoO_2(s). From the average oxidation potential and the $\Delta G_f°$ values for Po(s) and H_2O, the selected thermochemical values for PoO_2(s) are:

$$\Delta G_f°(PoO_2(s)) = -(192.1 \pm 3.3)\ \text{kJ mol}^{-1}$$

$$\Delta H_f°(PoO_2(s)) = -(251 \pm 5)\ \text{kJ mol}^{-1}$$

$$S_f°(PoO_2(s)) = (71 \pm 5)\ \text{J mol}^{-1}\text{K}^{-1}$$

where the uncertainties for the enthalpy and entropy have been assigned in the present review.

Eberhart and McDonald (1965) showed that a linear relationship often exists between the $\Delta G_f°$ and $\Delta H_f°$ values of the various physicochemical forms of an element. A plot of $\Delta H_f°$ against $\Delta G_f°$ is given in Fig. 3.5 for PoO_2(s) and the elemental polonium species discussed in Section 3.2.1. The line of best fit has a slope of 0.74 ± 0.02 [$r^2 = 0.9990$; $P \leq 0.001$;

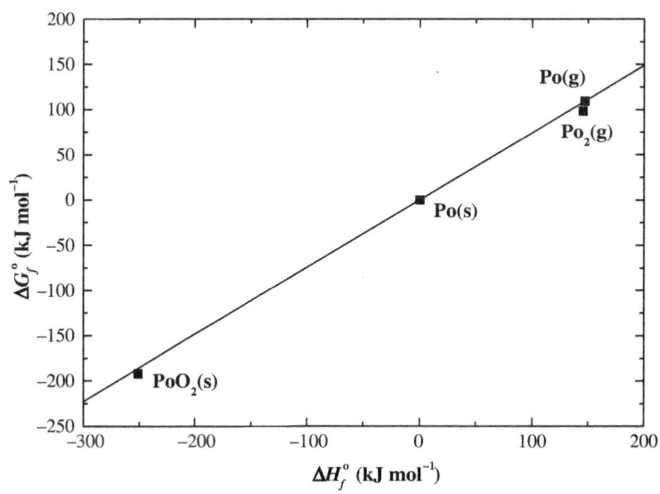

Figure 3.5 Plot of $\Delta H_f°$ versus $\Delta G_f°$ for PoO_2(s) and elemental polonium species.

$n = 4$] and must pass through the origin since, for the standard state (Po(s)), both ΔG_f° and ΔH_f° are 0.0 kJ mol^{-1}. It should be noted that the values of ΔG_f° and ΔH_f° as given by Zhdanov (1985) are the only data where both ΔG_f° and ΔH_f° have been published for polonium species, therefore limiting the number of points that can be used in Fig. 3.5.

3.2.5 H_2PoO_3(aq)

The solubility of PoO_2(s) in water can be described by the following equation.

$$PoO_2(s) + H_2O \rightleftharpoons H_2PoO_3(aq) \tag{3.42}$$

Bagnall et al. (1955) quote a solubility for PoO_2(s) of 0.075 mg L^{-1} in either water or excess alkali. The temperature was not given, and therefore ambient temperature is assumed (i.e., 18°C–30°C; this temperature range will be assumed where the temperature is not recorded for other species also (defined herein as ambient temperature). The average of the range, and the thermochemical data derived for it, is considered to be within the uncertainty limits of data that represent 25°C). From this, the solubility of PoO_2(s) was evaluated and an equilibrium constant, K°, of 3.57×10^{-7} was derived for Eq. (3.42). Use of this K°, and the ΔG_f° values for PoO_2(s) and H_2O, leads to the following Gibbs energy for H_2PoO_3(aq).

$$\Delta G_f^\circ(H_2PoO_3(aq)) = -(392.5 \pm 3.8)\ \text{kJ mol}^{-1}$$

The uncertainty has been derived by the present review using an assigned uncertainty for the log K° of Eq. (3.42) of 0.3.

3.2.6 PoO_3^{2-}

Bagnall and Freeman (1957) measured the solubility of polonium dioxide in KOH at 22°C [data in Bagnall (1957)]. A plot of the logarithm of [OH^-] versus the logarithm of polonium solubility (Fig. 3.6) has a slope of 2, and therefore the reaction can be described by Eq. (3.43). The measured solubility is consistent with that measured by Haring (1945) and Harlow (1947).

$$PoO_2(s) + 2OH^- \rightleftharpoons PoO_3^{2-} + H_2O \tag{3.43}$$

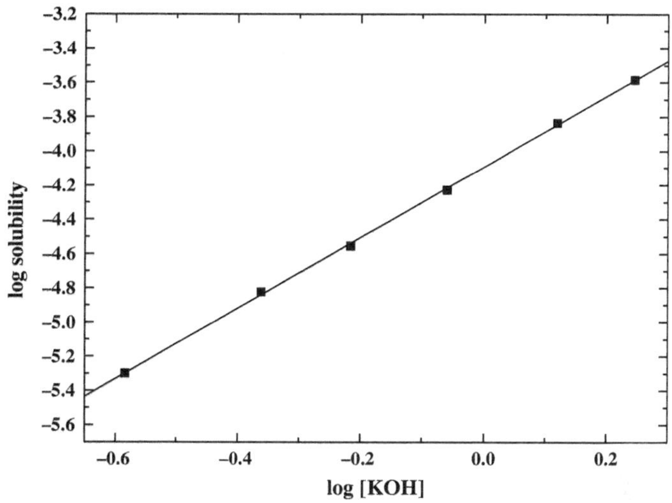

Figure 3.6 Plot of log [KOH] versus log solubility for $PoO_2(s)$ in hydroxide.

A value of $\log K°(PoO_3{}^{2-}) = -(4.43 \pm 0.02)$ has been derived for the solubility constant of Eq. (3.43) at zero ionic strength using the specific ion interaction theory (Grenthe et al., 1992) (Fig. 3.7) [$r^2 = 0.7765$; $P \leq 0.05$; $n = 6$]. The background to the specific ion interaction theory was given in Section 3.1.1.

Use of the derived log $K°$, and the $\Delta G_f°$ values for $PoO_2(s)$, OH^- and H_2O leads to the following Gibbs energy for PoO_3^{2-}.

$$\Delta G_f°(PoO_3^{2-}) = -(244.2 \pm 3.4)\ \text{kJ mol}^{-1}$$

Zhdanov (1985) gives a $\Delta G_f°$ value of -423 kJ mol^{-1} for PoO_3^{2-}; however, this appears to be erroneous since it would indicate that $PoO_2(s)$ is exceedingly soluble in base, a result that is inconsistent with the solubility measurements. Therefore use of the $\Delta G_f°$ value calculated from the data of Bagnall and Freeman (1957) for PoO_3^{2-} gives an oxidation potential ($E°$) for Eq. (3.44) of -0.031 V.

$$PoO_3^{2-} + 3H_2O + 4e^- \rightleftharpoons Po(s) + 6OH^- \quad (3.44)$$

The $E°$ value of -0.5 V for this reaction, as estimated by Latimer (1952) and used by Zhdanov (1985) to calculate the $\Delta G_f°$ value of -423 kJ mol^{-1}, has been assumed to be incorrect.

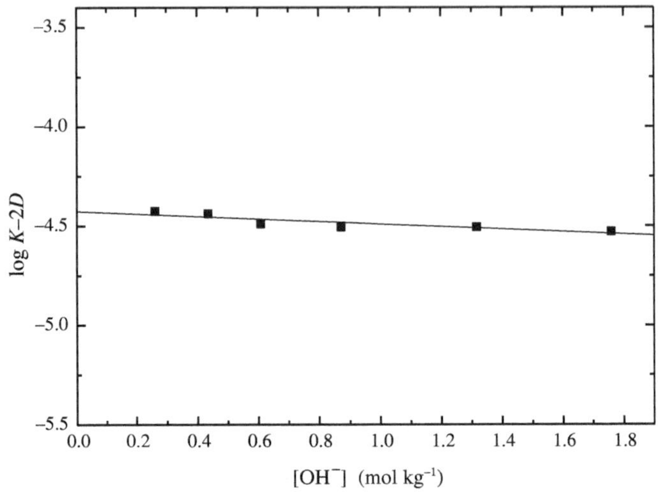

Figure 3.7 Plot of KOH molality versus log $K - 2D$ for PoO_2(s) in hydroxide to determine the log $K°$ of Eq. (3.43).

3.2.7 $HPoO_3^-$

Data for the formation of $HPoO_3^-$ have not been obtained; however, the data from both Bagnall and Freeman (1957) [see Bagnall (1957)] and Bagnall et al. (1955) can be used to estimate a log $K°$ for Eq. (3.45).

$$PoO_2(s) + OH^- \rightleftharpoons HPoO_3^- \tag{3.45}$$

As shown above, Bagnall et al. (1955) found that the solubility of PoO_2(s) in water and slightly alkaline solution is 0.075 mg L^{-1}. This indicates that the solubility of PoO_2(s) under these conditions is independent of pH [line 1 in Fig. 3.7; reaction (3.42)] with

$$\log K°(H_2PoO_3(aq)) = -(6.45 \pm 0.30)$$

where the uncertainty has been assigned in the present review. Additionally, Bagnall and Freeman (1957) demonstrated that a plot of the logarithm of polonium solubility versus the logarithm of hydroxide concentration, from 0.26 to 1.73 mol L^{-1}, has a slope of 2 (line 2 in Fig. 3.8) and a log $K°$ of -4.43. This indicates that two moles of hydroxide are consumed for every one mole of polonium solubilized. The intermediate condition, represented by reaction (3.45), will therefore have an intermediate log $K°$ value. Line 3 in Fig. 3.8 is used to represent the chemical behavior of Eq. (3.45), i.e., it has a slope of 1 indicating that for every mole of polonium solubilized one mole of hydroxide is consumed.

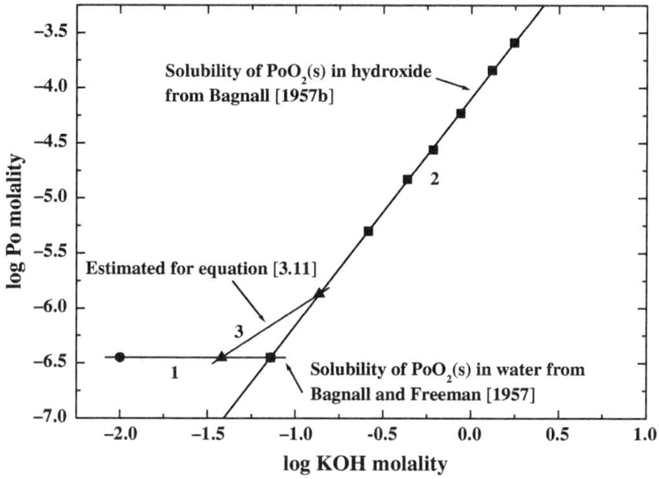

Figure 3.8 Plot of solubility of PoO_2(s) in hydroxide and water to determine the log $K°$ of Eq. (3.45).

The projection of this line from line 2 is at a molality of 0.13 which represents the midpoint (i.e., the point which has the highest uncertainty) between the lowest concentration used by Bagnall and Freeman (1957) (i.e., 0.26 mol L^{-1}) and zero (negligible OH^- concentration).

The intercept of line 3 in Fig. 3.8 is the log $K°$ for Eq. (3.45).

$$\log K°\left(HPoO_3^-\right) = -(5.01 \pm 0.51)$$

The large uncertainty is due to the choice of the midpoint for the assessment of the solubility constant.

From this log $K°$, and the $\Delta G_f°$ values for PoO_2(s) and OH^-, a Gibbs energy of -320.8 ± 4.4 kJ mol^{-1} can be calculated for $HPoO_3^-$. This analysis may lead to a Gibbs energy that is somewhat speculative, but $HSeO_3^-$ and $HTeO_3^-$ are known species for selenium and tellurium, respectively. However, due to the speculative nature of the analysis, a second methodology for estimating the Gibbs energy of $HPoO_3^-$ was derived.

A linear free energy relationship has been found to exist between the Gibbs energy values of tellurium species and compounds and the equivalent species and compounds of polonium. The species and compounds include H_2X(aq), HX^-, X^{2-}, XO_2(s), $XOOH^+$, H_2XO_3(aq), HXO_3^-, and XO_3^{2-}, where X is either tellurium or polonium. The data for tellurium are listed in Tables 3.2 and 3.3, or Appendix 1 (Panson, 1963; Zhdanov, 1985; McPhail, 1995). The data for polonium are listed in the

Table 3.3 Calculation of $\Delta G_f°$ values of HX^- for O, S, Se, and Te [log $K°$ values from Högfeldt (1982); Brown and Ekberg (2016)].

Chalcogen	$\Delta G_f°$ (H_2X)[a] Data from Latimer (1952)	$\Delta G_f°$ (H_2X)[a] Data from more recent literature	log $K°$ $H_2X \leftrightarrow HX^- + H^+$	$\Delta G_f°$ (HX^-)[b] Data from Latimer (1952)	$\Delta G_f°$ (HX^-)[b] Data from more recent literature
O			− 13.994		− 157.3
S	− 27.0		− 6.99	12.2	
Se	77.2	21.4	− 3.81	98.9	43.5
Te	138.9	98.4	− 2.64	154.0	113.5

[a]Derived using the H_2X/X oxidation potentials in Table 3.2.
[b]Derived using the $\Delta G_f°$ values for H_2X.

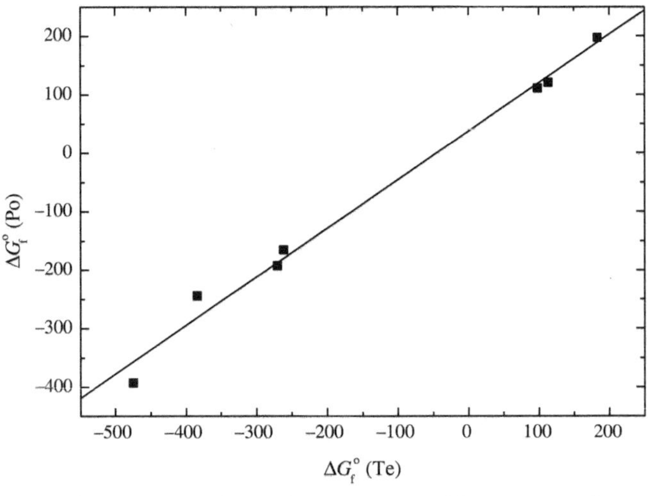

Figure 3.9 Linear free energy relationship between Gibbs energy values of tellurium and polonium species and compounds.

relevant sections of this chapter. The linear free energy relationship is illustrated in Fig. 3.9. The line of best fit to the data has a slope of 0.83 ± 0.04 and an intercept of 37.1 ± 11.0 [$r^2 = 0.7765$; $P \leq 0.05$; $n = 6$]. The Gibbs energy for $HTeO_3^-$ is $\Delta G_f° = -438.2$ kJ mol^{-1} (McPhail, 1995) and from this value and the line of best fit the Gibbs energy calculated for $HPoO_3^-$ is -326.1 kJ mol^{-1}.

The Gibbs energy for $HPoO_3^-$ is taken as the weighted average of the two values determined:

$$\Delta G_f^\circ(HPoO_3^-) = -(321.6 \pm 4.1)\ \text{kJ mol}^{-1}$$

where the uncertainty is taken from the weighted average of the two values. An uncertainty of 10 kJ mol^{-1} was assigned to the second Gibbs energy derived for $HPoO_3^-$.

The proton association reactions for $HPoO_3^-$ and PoO_3^{2-} can be described by Eqs. (3.46) and (3.47).

$$HPoO_3^- + H^+ \rightleftharpoons H_2PoO_3(aq) \tag{3.46}$$

$$PoO_3^{2-} + H^+ \rightleftharpoons HPoO_3^- \tag{3.47}$$

Use of the ΔG_f° values for $HPoO_3^-$, $H_2PoO_3(aq)$ and PoO_3^{2-} gives:

$$\log K^\circ(H_2PoO_3(aq)) = 12.41 \pm 0.30$$

$$\log K^\circ(HPoO_3^-) = 13.56 \pm 0.30$$

for Eqs. (3.46) and (3.47), respectively. The uncertainties have been assigned by this review. For the two species, $HPoO_3^-$ can be equivalently written as $PoO(OH)_3^-$ and PoO_3^{2-} as $PoO(OH)_4^{2-}$, the third and fourth hydrolyzed species of the tetravalent polonium(IV) ion PoO^{2+}. Moreover, $H_2PoO_3(aq)$ can also be equivalently written as the second hydrolysis species $PoO(OH)_2(aq)$.

3.2.8 PoO^{2+} and $PoOOH^+$

Hataye et al. (1981) investigated the hydrolysis of polonium(IV) ions by solvent extraction using dithizone/carbon tetrachloride solutions. For Eq. (3.48), a log K of 1.12 was obtained in 1.0 mol L^{-1} (H, Na)ClO_4 at ambient temperature.

$$PoO(OH)_2(aq) + H^+ \rightleftharpoons PoO(OH)^+ + H_2O \tag{3.48}$$

Given the equivalence of $PoO(OH)_2(aq)$ and $H_2PoO_3(aq)$, Eq. (3.48) may also be written as:

$$H_2PoO_3(aq) + H^+ \rightleftharpoons PoO(OH)^+ + H_2O \tag{3.49}$$

Since there is no change in the sum of the charge of the products and reactants, there will be little difference between the measured log K and the log K° (zero ionic strength).

Starik et al. (1964a) also investigated the hydrolysis of polonium(IV) using solvent extraction at ambient temperature. Ampelogova (1973) reanalyzed this work and found evidence for the species PoO^{2+}, $PoO(OH)^{+}$ and $PoO(OH)_2(aq)$ in 0.1 mol L^{-1} (Na, H)ClO_4. For Eqs. (3.50) and (3.51), the overall stability constants (log β_1 and log β_2) derived were -1.06 and -3.26, respectively.

$$PoO^{2+} + H_2O \rightleftharpoons PoO(OH)^{+} + H^{+} \quad (3.50)$$

$$PoO^{2+} + 2H_2O \rightleftharpoons PoO(OH)_2(aq) + 2H^{+} \quad (3.51)$$

When converted to zero ionic strength using the Davies equation (Davies, 1962), the values (log $\beta_1°$ and log $\beta_2°$) become -0.85 and -3.06, respectively, and therefore the stepwise stability constant (log $K_2°$) for Eq. (3.52) is -2.21.

$$PoO(OH)^{+} + H_2O \rightleftharpoons PoO(OH)_2(aq) + H^{+} \quad (3.52)$$

Since this reaction is the reverse of that given by Hataye et al. (1981), the average stability constant for Eq. (3.49) is:

$$\log K°\left(PoO(OH)_2(aq)\right) = 1.7 \pm 0.6$$

where the uncertainty is chosen to span the range in the two independent values. Use of this log $K°$ and the $\Delta G_f°$ values for $H_2PoO_3(aq)$ and H_2O gives:

$$\Delta G_f°\left(PoO(OH)^{+}\right) = -(164.8 \pm 4.9)\ \text{kJ mol}^{-1}$$

In turn, a Gibbs energy for PoO^{2+} has been calculated using -0.85 for the log $\beta_1°$ of Eq. (3.50) and the $\Delta G_f°$ values for $PoO(OH)^{+}$ and H_2O.

$$\Delta G_f°\left(PoO^{2+}\right) = (67.6 \pm 5.2)\ \text{kJ mol}^{-1}$$

The uncertainty for log $\beta_1°$ of 0.3 has been assigned by the present review.

Koch and Schmidt (1963) studied the hydrolysis of polonium(IV) using an ion exchange technique. They postulated that Po^{4+} transitioned to $PoOH^{3+}$ over a pH range of 2–3.3 and they quoted a log K for this reaction of -3.4. Further, $PoOH^{3+}$ changed to $Po(OH)_2^{2+}$ over the pH range of 3.3–5 with an overall log $\beta = -8.2$, $Po(OH)_2^{2+}$ transitioned to $Po(OH)_3^{+}$ in the pH range of 5–7, and finally, $Po(OH)_3^{+}$ changed to Po

$(OH)_4(aq)$ at a pH greater than 7. In a later study, Koch and Falkenburg (1967) demonstrated that below a pH of about 3 that a divalent ion (most probably PoO^{2+}) was present in the aqueous phase. A neutral species was found to exist above a pH of 3.2 [this species is most likely $H_2PoO_3(aq)$]. The results from this latter work are inconsistent with the earlier work of Koch and Schmidt (1963), but are consistent with the observations of Hataye et al. (1981) and Starik et al. (1964a) [as recalculated by Ampelogova (1973)]. Koch and Falkenburg (1967) did not determine stability constants for any hydrolysis species.

3.2.9 $PoO_3(s)$

In acid solution, $PoO_3(s)$ reacts to form $PoO_2(s)$ according to the following equation.

$$PoO_3(s) + 2H^+ + 2e^- \rightleftharpoons PoO_2(s) + H_2O \tag{3.53}$$

A value of 1.509 V for the $E°$ of the $PoO_3(s)/PoO_2(s)$ couple and -138 kJ mol^{-1} for the $\Delta G_f°$ of $PoO_3(s)$ were obtained from Zhdanov (1985) and Latimer (1952), respectively. Use of this $E°$ value and the $\Delta G_f°$ values for $PoO_2(s)$ and H_2O gives a calculated $\Delta G_f°$ of -138.1 kJ mol^{-1} for $PoO_3(s)$. This is in excellent agreement with that of Latimer (1952), who had estimated his value.

For Eq. (3.54), an $E°$ value of 0.55 V was determined by Haissinsky (1946) at ambient temperature.

$$PoO_3(s) + 2e^- \rightleftharpoons PoO_3^{2-} \tag{3.54}$$

From this $E°$ value and the $\Delta G_f°$ for PoO_3^{2-}, also leads to a $\Delta G_f°$ of -138.1 kJ mol^{-1} for $PoO_3(s)$. The calculated uncertainty from both $E°$ values for the Gibbs energy for $PoO_3(s)$ is 4.4 kJ mol^{-1}. Thus the selected Gibbs energy for $PoO_3(s)$ is the weighted average of the two values determined (assuming they are independent).

$$\Delta G_f°(PoO_3(s)) = -(138.1 \pm 3.1)\ \text{kJ mol}^{-1}$$

3.2.10 $H_2Po(aq)$ and HPo^-

As indicated in Section 3.2.2, Brown (2001) used the data of Latimer (1952) and the methodology of Kapustinskii (1948) to determine thermochemical data for $H_2Po(aq)$. However, the more recent thermochemical data for both selenium and tellurium invalidated the use of the

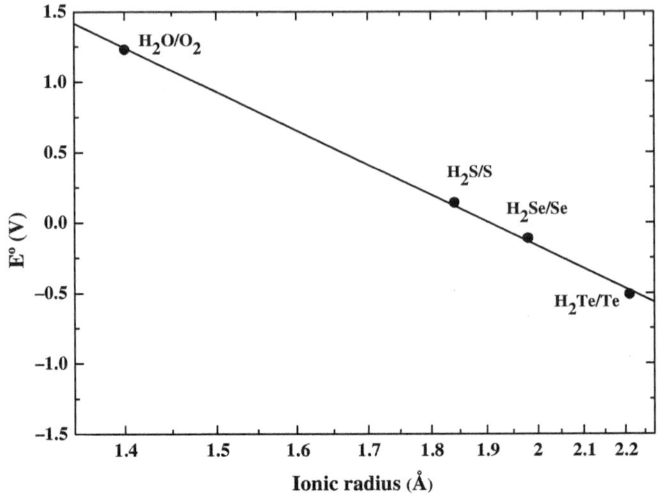

Figure 3.10 Plot of $1/r$ versus $E°$ for the H_2X/X oxidation couple (X is O, S, Se, or Te).

methodology of Kapustinskii. Consequently, a similar methodology as was used to derive an oxidation potential for Po^{2-}, was also used for the oxidation potential of $H_2Po(aq)$ using the data that are listed in Table 3.3.

Fig. 3.10 contains a plot of $1/r$ of the respective chalcogen versus $E°$ for the H_2X/X oxidation couple of O, S, Se, and Te. The intercept and slope of the line of best fit are -3.45 ± 0.17 and 6.57 ± 0.10, respectively [$r^2 = 0.9987$; $P \leq 0.001$; $n = 4$]. The oxidation potential for polonium determined from these linear relationships, and the ionic radius of 2.284 Å for Po^{2-} (Bagnall, 1983), is -0.57 V for the $H_2Po(aq)/Po(s)$ oxidation couple. Again, this value is substantially more positive than that estimated by Latimer (1952) as was also found for the more recent oxidation potentials of the $H_2Se(aq)/Se(s)$ and $H_2Te(aq)/Te(s)$ oxidation couples (see Table 3.3).

From the estimated oxidation potential ($E°$) of -0.57 V for the $H_2Po(aq)/Po(s)$ oxidation couple, the Gibbs energy calculated for $H_2Po(aq)$ is:

$$\Delta G_f°(H_2Po(aq)) = (110.8 \pm 7.7)\ \text{kJ mol}^{-1}$$

The uncertainty has been assigned by the present review on the basis of an uncertainty assigned to the derived $E°$ of 0.04 V.

Brown (2001) used the methodology of Kapustinskii (1948) to calculate the $\Delta G_f°$ value for HPo^- using the $\Delta G_f°$ values of HX^- for S, Se,

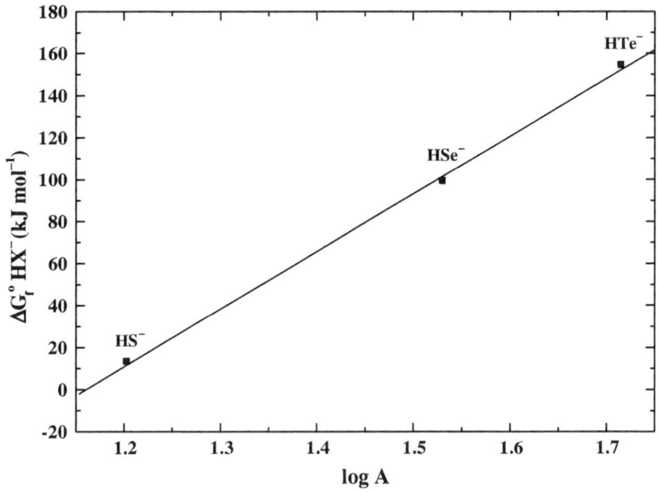

Figure 3.11 Plot of log A versus $\Delta G_f°$ of HX^-(X is S, Se, or Te).

and Te given in Table 3.3. The $\Delta G_f°$ value for OH^- is also given in the table (Baes and Mesmer, 1976; Sillén and Martell, 1964, 1971; Brown and Ekberg, 2016).

A plot of log A versus $\Delta G_f°$ is shown in Fig. 3.11. The line of best fit has an intercept of -318 ± 14 and a slope of 274 ± 9 [$r^2 = 0.9989$; $P \leq 0.05$; $n = 3$]. From these data, Brown (2001) calculated a $\Delta G_f°$ of 209.3 kJ mol^{-1} for HPo^-. Use of this value and the $\Delta G_f°$ values for H_2Po(aq) and Po^{2-} gives log $K°$ values for Eqs. (3.55) and (3.56) of -0.49 and -11.33, respectively.

$$H_2Po(aq) \rightleftharpoons HPo^- + H^+ \tag{3.55}$$

$$HPo^- \rightleftharpoons Po^{2-} + H^+ \tag{3.56}$$

As indicated above, the methodology used by Kapustinskii (1948) is not valid. The data have been reinterpreted using the relationship between $\Delta G_f°$ and $1/r$, illustrated in Fig. 3.12. The intercept and slope of the line of best fit are 564 ± 25 and -1012 ± 44, respectively [$r^2 = 0.9963$; $P \leq 0.001$; $n = 4$]. From these data, the Gibbs energy for HPo^- is calculated.

$$\Delta G_f°(HPo^-) = (120.4 \pm 7.0)\ \text{kJ mol}^{-1}$$

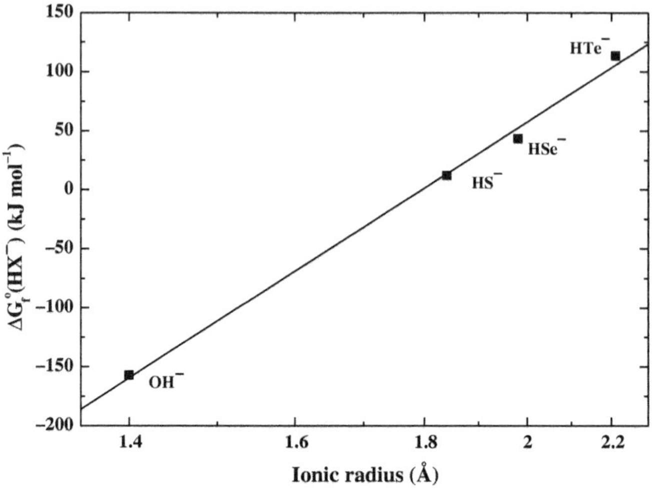

Figure 3.12 Plot of 1/*r* versus ΔG° of HX^- species (X is O, S, Se, or Te).

Use of this value and the ΔG_f° values for $H_2Po(aq)$ and Po^{2-} gives:

$$\log K^\circ(HPo^-) = -1.69$$

$$\log K^\circ\left(Po^{2-}\right) = -13.44$$

for reactions (3.55) and (3.56), respectively. The uncertainty for the Gibbs energy has been assigned by the present review.

3.2.11 $PoCl_2(s)$

It has been shown (Fig. 3.5) that a correlation exists between ΔH_f° and ΔG_f°, and therefore it follows that a similar relationship should exist between ΔS_f° and ΔG_f°. Values of ΔS_f° can be calculated from values of S_f° using Eq. (3.18). The relationship between ΔS_f° and ΔG_f° for the polonium species used in Fig. 3.5 [Po(s), Po(g), $Po_2(g)$, and $PoO_2(s)$], is shown in Fig. 3.13. Again, since the values of the standard state, Po(s), are, by definition, both zero, the line of best fit must pass through the origin. The slope of the line is 1.15 ± 0.12 [$r^2 = 0.9844$; $P \leq 0.01$; $n = 4$]. There is clearly more uncertainty in the relationship shown in Fig. 3.13 than was apparent in that of the relationship shown in Fig. 3.5.

Zhdanov (1985) gives a value of $130\,J\,mol^{-1}\,K^{-1}$ for the S_f° of $PoCl_2(s)$ from which a ΔS_f° of $-155.9 \pm 2.0\,J\,mol^{-1}\,K^{-1}$ is obtained.

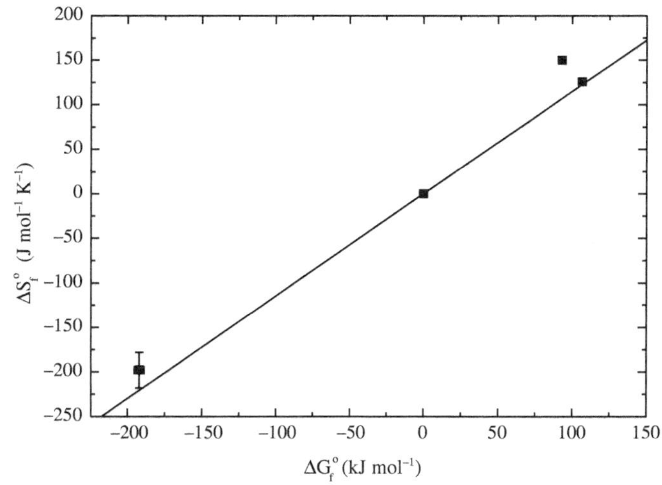

Figure 3.13 Plot of $\Delta S_f°$ versus $\Delta G_f°$ for Po(s), Po(g), $Po_2(g)$ and $PoO_2(s)$.

From this $\Delta S_f°$ and the line of best fit given in Fig. 3.13, the calculated value for the Gibbs energy of $PoCl_2(s)$ is:

$$\Delta G_f°(PoCl_2(s)) = -(136 \pm 14)\ \text{kJ mol}^{-1}$$

Due to the relatively large uncertainty indicated by Fig. 3.13, a large uncertainty has been assigned by this review to the Gibbs energy of $PoCl_2(s)$.

3.2.12 $PoCl_4(s)$

For metal ions possessing an inert pair of electrons, namely, In^+, Tl^+, Sn^{2+}, Pb^{2+}, Sb^{3+}, Bi^{3+}, Te^{4+}, and Po^{4+}, both the $\Delta G_f°$ and $\Delta H_f°$ values of the metal halides in the same group of the Periodic Table are similar. For example, the $\Delta G_f°$ and $\Delta H_f°$ values of $SnBr_2(s)$ are −248.9 and −266.1 kJ mol^{-1}, respectively, which are similar to those for $PbBr_2(s)$, namely, −260.4 and −277.0 kJ mol^{-1}. A plot of $\Delta G_f°$ values for the MX_b species of In^+, Sn^{2+}, Sb^{3+}, and Te^{4+} (for these ions, n, the principal quantum number, equals five and X is Cl, Br, or I) against the respective values for the MX_b species of Tl^+, Pb^{2+}, Bi^{3+}, and Po^{4+} ($n = 6$) is linear. Similarly, a plot of $\Delta H_f°$ values is also linear. Table 3.4 summarizes the $\Delta G_f°$ and $\Delta H_f°$ values taken from Bard et al. (1985) and this work. The plots are shown in Figs. 3.14 and 3.15, respectively, and will allow thermodynamic data for the PoX_b species to be determined when the corresponding data for tellurium are known.

Table 3.4 $\Delta G_f°$ and $\Delta H_f°$ values for various metal halide species (MX_b; $n = 5$ or 6) for $X = Cl^-$, Br^-, or I^- [values from Bard et al. (1985) except $PoCl_6^{2-}$ as determined in this work].

Metal halide	n	$\Delta G_f°$ (kJ mol^{-1})	$\Delta H_f°$ (kJ mol^{-1})
InCl	5		−186
InBr		−169	−175
InI		−120	−116
$SbCl_3$		−323.72	−382.27
$SbBr_3$		−239.3	−259.4
$SnCl_2$		−302.1	−349.8
$SnBr_2$		−248.9	−266.1
$TeCl_6^{2-}$		−574.9	
TlCl	6		−204.1
TlBr		−167	−173
TlI		−125.39	−123.8
$BiCl_3$		−314.6	−379.1
$BiBr_3$		−247.7	−276.1
$PbCl_2$		−314.0	−359.20
$PbBr_2$		−260.41	−277.02
$PoCl_6^{2-}$		−565.6	

The line of best fit in Fig. 3.14 has an intercept of -12 ± 7 and a slope of 0.97 ± 0.02 [$r^2 = 0.9970$; $P \leq 0.001$; $n = 7$] and that in Fig. 3.15, an intercept of -13 ± 10 and a slope of 0.98 ± 0.04 [$r^2 = 0.9927$; $P \leq 0.001$; $n = 7$].

Zhdanov (1985) quotes values of -237.2 and -326.4 kJ mol^{-1} for the $\Delta G_f°$ and $\Delta H_f°$ of $TeCl_4$(s), respectively. The calculated values for the $\Delta G_f°$ and $\Delta H_f°$ of $PoCl_4$(s), using these data and the regression equations from Figs. 3.14 and 3.15, are -241.1 and -333.1 kJ mol^{-1}, respectively. Zhdanov (1985) also quotes an $S_f°$ value for $PoCl_4$(s) of 197 J mol^{-1} K^{-1}. All of these thermochemical values are consistent with the values of -241.8 kJ mol^{-1}, -334.7 kJ mol^{-1}, and 196.6 J mol^{-1} K^{-1}, respectively, reported by Ruzinov and Giljanickij (1975). From the $\Delta S_f°$ calculated from this $S_f°$ (Eq. (3.18)) and the $\Delta H_f°$ value of -333.1 kJ mol^{-1}, a $\Delta G_f°$ value of -240.1 kJ mol^{-1} for $PoCl_4$(s) is obtained using Eq. (3.6). Similarly, using -333.1 kJ mol^{-1} for the $\Delta H_f°$ and the regression equation from Fig. 3.5, a $\Delta G_f°$ value of -247.7 kJ mol^{-1} has been calculated. The average of the three $\Delta G_f°$ values is: $\Delta G_f°(PoCl_4(s)) = -(243.0 \pm 4.7)$ kJ mol^{-1} where the uncertainty has been chosen to span the range in the three Gibbs energy values.

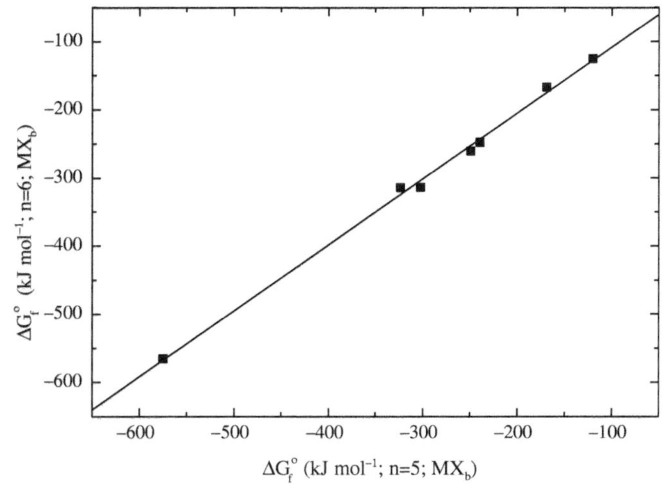

Figure 3.14 Plot of $\Delta G_f°$ MX_b (n = 5) versus $\Delta G_f°$ MX_b (n = 6) for X = Cl^-, Br^- or I^-.

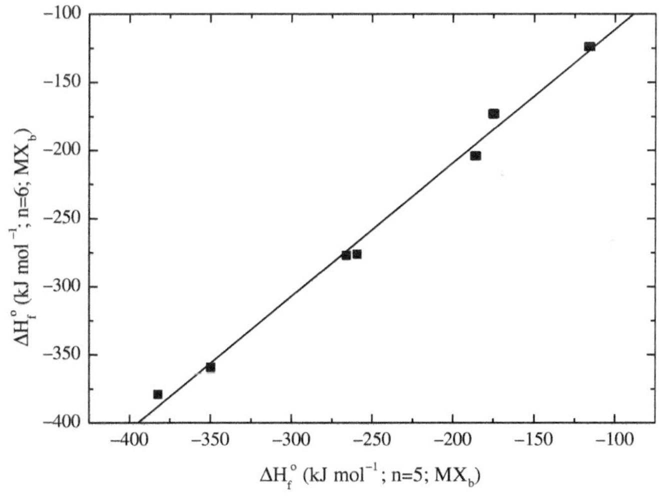

Figure 3.15 Plot of $\Delta H_f°$ MX_b (n = 5) versus $\Delta H_f°$ MX_b (n = 6) for X = Cl^-, Br^- or I^-.

3.2.13 $PoCl_4^{2-}$

For Eq. (3.57), Eichelberger et al. (1965) determined oxidation potentials of 0.417, 0.387, 0.367, 0.342, and 0.297 V in 1.0, 1.5, 2.0, 3.0, and 4.0 mol L^{-1} HCl, respectively, at 25°C. These values are in good

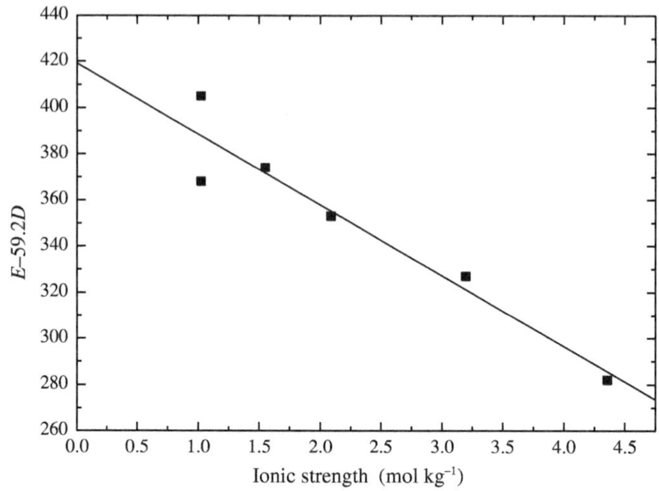

Figure 3.16 Plot of ionic strength versus $E - 59.2D$ for the $PoCl_4^{2-}$/Po oxidation couple.

agreement with that of 0.38 V, as determined by Bagnall and Freeman (1956a) in 1 mol L^{-1} HCl at 22°C.

$$PoCl_4^{2-} + 2e^- \rightleftharpoons Po(s) + 4Cl^- \tag{3.57}$$

Use of these E values, and correcting to zero ionic strength using the specific ion interaction theory (Grenthe et al., 1992), allows an $E°$ value for the $PoCl_4^{2-}$/Po(s) oxidation couple to be determined from a plot of ionic strength against $E - 59.2D$ (Fig. 3.16).

The graph shows good agreement between the parameters and gives an intercept ($E°$) of 0.419 ± 0.012 V [$r^2 = 0.9179$; $P \leq 0.01$; $n = 6$]. Use of this $E°$ and the $\Delta G_f°$ values for Po(s) and Cl^- leads to a Gibbs energy for $PoCl_4^{2-}$ of:

$$\Delta G_f°(PoCl_4^{2-}) = -(444.0 \pm 2.3)\ \text{kJ mol}^{-1}.$$

In turn, use of this $\Delta G_f°$ value and the $\Delta G_f°$ values for Po^{2+} and Cl^- gives:

$$\log \beta_4°(PoCl_4^{2-}) = 7.58$$

for Eq. (3.58).

$$Po^{2+} + 4Cl^- \rightleftharpoons PoCl_4^{2-} \tag{3.58}$$

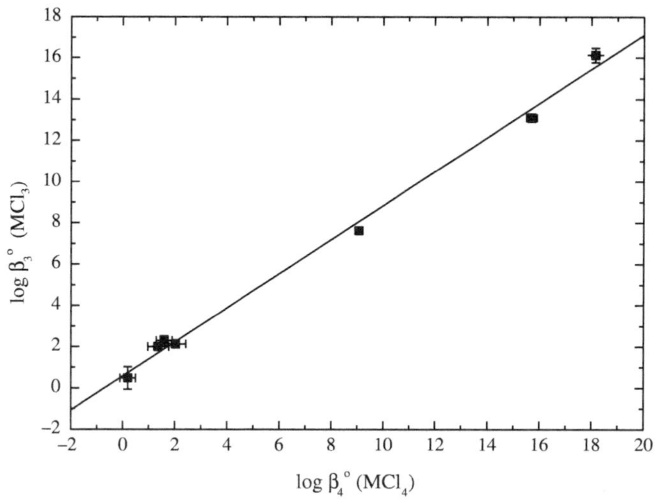

Figure 3.17 Linear free energy relationship between the stability constants of MCl_4 and MCl_3 complexes.

3.2.14 $PoCl^+$, $PoCl_2$(aq), and $PoCl_3^-$

By studying the relationship between the overall stability constants for metal chloride species, it may be possible to determine the corresponding polonium values. The linear free energy relationships between $MCl_n^{(z-n)+}$ and $MCl_{n-1}^{(z-n+1)+}$ species are represented graphically in Figs. 3.15–3.17 for the averaged stability constant data for various metal ion chloride species obtained from the literature, which are summarized in Table 3.5.

The general equation describing the formation of the chloride species discussed above is:

$$Po^{2+} + nCl^- \rightleftharpoons PoCl_4^{(2-n)} \tag{3.59}$$

The line of best fit in Fig. 3.17 has an intercept of 0.6 ± 0.2 and a slope of 0.83 ± 0.02 [$r^2 = 0.9955$; $P \leq 0.001$; $n = 7$]. Similarly, Figs. 3.18 and 3.19 have intercepts and slopes of 0.7 ± 0.2 and 0.78 ± 0.03 [$r^2 = 0.9906$; $P \leq 0.001$; $n = 8$] and 0.52 ± 0.07 and 0.51 ± 0.01 [$r^2 = 0.9903$; $P \leq 0.001$; $n = 18$], respectively.

From the previously determined value of 7.58 for the log $\beta_4°$ of $PoCl_4^{2-}$ and the line of best fit from Fig. 3.17, the log $\beta_3°$ for $PoCl_3^-$ can be calculated. In turn, use of this log $\beta_3°$ and the line of best fit from Fig. 3.18 will lead to the log $\beta_2°$ for $PoCl_2$(aq). Finally, use of this latter

Table 3.5 log β_n° Values of metal chloride complexes at 25°C and zero ionic strength.

Metal ion	Species[a]			
	$MCl^{(z-1)}$	$MCl_2^{(z-2)}$	$MCl_3^{(z-3)}$	$MCl_4^{(z-4)}$
Ag(I)	3.04	5.14		
Am(III)	0.24	− 0.74		
Bi(III)	3.65	5.85	7.62	9.06
Cd(II)	1.98	2.64	2.30	1.65
Cr(III)	0.60	− 0.11		
Cu(II)	0.83	0.60		
Fe(III)	1.52	2.22	1.02	
Hg(II)	7.31	14.00		
Pb(II)	1.50	2.10	2.00	1.03
Pd(II)	6.05	10.65	13.10	15.40
PuO_2(II)	0.23	− 1.15		
Sn(II)	1.52	2.17	2.13	2.03
Sn(IV)	3.19	5.95		
Tl(I)	0.57	− 0.11		
Tl(III)	7.93	13.54	16.13	18.15
UO_2(II)	0.17	− 1.10		
Zn(II)	0.40	0.69	0.48	0.54
Zr(IV)	1.59	2.17		

[a]z = the ionic charge of the metal.
Source: Data sources listed in Appendix 1.

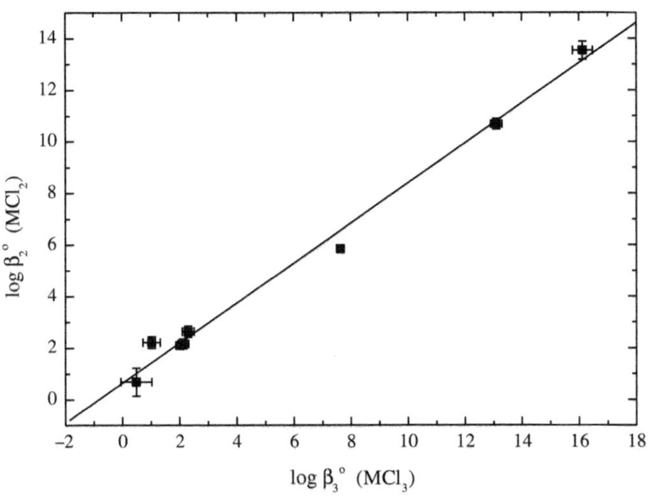

Figure 3.18 Linear free energy relationship between the stability constants of MCl_3 and MCl_2 complexes.

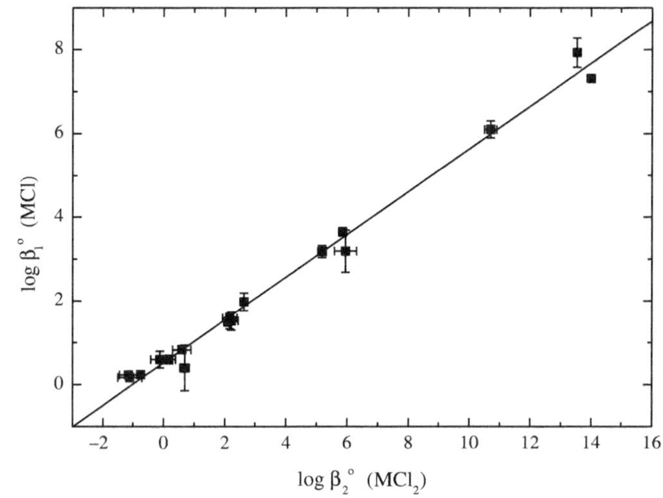

Figure 3.19 Linear free energy relationship between the stability constants of MCl_2 and MCl complexes.

stability constant and the line of best fit from Fig. 3.19, will permit calculation of log $\beta_1°$ for $PoCl^+$. The derived stability constants based on Eq. (3.59) are:

$$\log \beta_1°\left(PoCl^+, n = 1\right) = 3.56 \pm 0.20$$

$$\log \beta_2°(PoCl_2(aq), n = 2) = 5.95 \pm 0.20$$

$$\log \beta_3°\left(PoCl_3^-, n = 3\right) = 6.84 \pm 0.20$$

The uncertainties have been assigned by the present review.

The corresponding $\Delta G_f°$ values subsequently derived for $PoCl_3^-$, $PoCl_2(aq)$, and $PoCl^+$ using the $\Delta G_f°$ values for Po^{2+} and Cl^- are:

$$\Delta G_f°\left(PoCl_3^-\right) = -(308.5 \pm 8.4)\ \text{kJ mol}^{-1}.$$

$$\Delta G_f°(PoCl_2(aq)) = -(172.2 \pm 8.4)\ \text{kJ mol}^{-1}.$$

$$\Delta G_f°\left(PoCl^+\right) = -(27.4 \pm 8.4)\ \text{kJ mol}^{-1}.$$

3.2.15 $PoCl_6^{2-}$

For Eq. (3.60), Eichelberger et al. (1965) determined electrode potentials of 0.717 and 0.702 V in 1.0 and 1.5 mol L^{-1} HCl, respectively, at 25°C.

$$PoCl_6^{2-} + 2e^- \rightleftharpoons PoCl_4^{2-} + 2Cl^- \tag{3.60}$$

The former value is in excellent agreement with the 0.72 V measured by Bagnall and Freeman (1956a) in 1 mol L^{-1} HCl at 22°C. Power (1949a, 1949b) measured a potential of 0.38 V for the Po^{IV}/Po^{II} couple in 4.7 mol L^{-1} HCl using a Ag/AgCl electrode, which is equivalent to a potential of 0.582 V versus a standard hydrogen electrode [data in Moyer (1956); measured at ambient temperature]. From these values, and using the specific ion interaction theory (Grenthe et al., 1992), a plot of ionic strength against $E - 118.4D$ (Fig. 3.18) gives a straight line with an intercept equivalent to $E°$.

From the line of best fit for Fig. 3.20, an $E°$ value of 0.730 ± 0.001 V was obtained [$r^2 = 0.9997$; $P \leq 0.001$; $n = 4$]. Use of this $E°$ and the $\Delta G_f°$ values for $PoCl_4^{2-}$ and Cl^- leads to a Gibbs energy for $PoCl_6^{2-}$ of:

$$\Delta G_f°\left(PoCl_6^{2-}\right) = -(565.6 \pm 0.5)\ \text{kJ mol}^{-1}$$

For Eq. (3.61),

$$PoO^{2+} + 6Cl^- + 2H^+ \rightleftharpoons PoCl_6^{2-} + H_2O \tag{3.61}$$

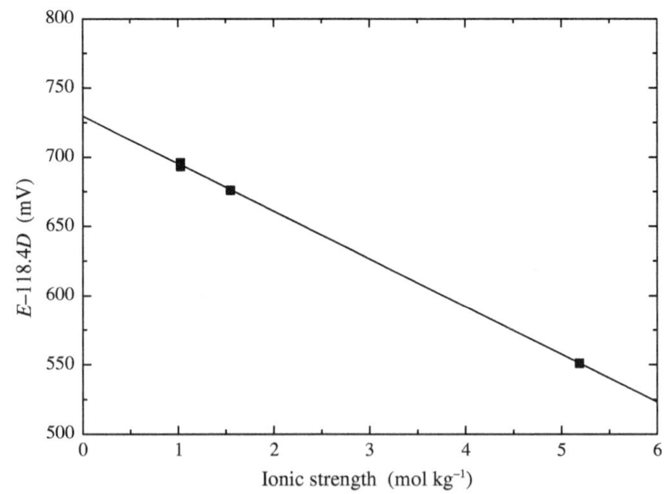

Figure 3.20 Plot of ionic strength versus $E - 118.4D$ for the $PoCl_6^{2-}$/Po oxidation couple.

a stability constant of:

$$\log K^{\circ}\left(PoCl_6^{2-}\right) = 14.55$$

has been calculated from the ΔG_f° values for $PoCl_6^{2}$, PoO^{2+}, H_2O, and Cl^-. This stability constant is consistent with that given for Eq. (3.61) by Bagnall and Freeman (1956a) of $\log K = 14$ for 1 mol L^{-1} HCl and 22°C.

The complexation of polonium(IV) by chloride was studied by Starik et al. (1964b) and Starik and Ampelogova (1965) at ambient temperature using 1 and 4–6 mol L^{-1} H(Cl, ClO_4), respectively. In both studies, they postulated the formation of six species, from $PoCl^{3+}$ to $PoCl_6^{2-}$, suggesting that the reacting polonium species was Po^{4+}. However, there is substantial evidence that at low concentrations of chloride, the polonium species in solution should be PoO^{2+}. The stability constant obtained by Starik et al. (1964b) for the formation of $PoCl_6^{2-}$ was 11.6 (log *K*), more than two orders of magnitude less than that obtained by Bagnall and Freeman (1956a) at the same ionic strength. The stability constant obtained by Starik and Ampelogova (1965) in the higher ionic strength was similar to that obtained in their earlier paper ($\log K = 11.9$). Given issues associated with the data of Starik and coworkers, with respect to both the species that form in the aqueous solution as well as in the organic phase, the data of these authors are not accepted in the present work (see Section 3.2.17).

Younes (2013) studied polonium speciation in chloride media with solvent extraction. In this work, confirmation was given for the formation of the divalent anion $PoCl_6^{2-}$. It was also suggested that two other polonium species, $PoCl_4(aq)$ and $PoOCl_2(aq)$ [called $Po(OH)_2Cl_2$ by Younes (2013)], could be in equilibrium with $PoCl_6^{2-}$ depending on the chloride concentration. The following two reactions were proposed:

$$PoCl_4(aq) + 2Cl^- \rightleftharpoons PoCl_6^{2-} \quad (3.62)$$

$$PoOCl_2(aq) + 4Cl^- + 2H^+ \rightleftharpoons PoCl_6^{2-} + 2H_2O \quad (3.63)$$

where these reactions had reported stability constants of $\log K = 4.6 \pm 0.2$ and 2.7 ± 0.8, respectively. From these data, and that given above for $PoCl_6^{2-}$, thermochemical data could be derived for the two neutral species, $PoCl_4(aq)$ and $PoOCl_2(aq)$. However, this has not been done, because in a subsequent publication (Younes et al., 2017), although the three aqueous species were described, the thermochemical data for their interaction were not included. This exclusion indicates that the authors of the work were not confident that the $PoCl_4$(aq) and $PoOCl_2$(aq) species existed in aqueous solution. However, this is not to say that they do not, particularly given their similarity to other polonium chloride species that have acceptable thermochemical data, $PoOHCl_4^-$ and $PoO(OH)Cl_2^-$ (see Sections 3.2.16 and 3.2.17). Given the similarity in the stoichiometry of these species, it may be possible that Younes (2013) misinterpreted the two former species for the latter two species (particularly since they only differ from the latter species by inclusion of a hydroxide molecule). As a consequence of this uncertainty, the species of Younes (2013) have not been accepted in the present review.

For Eq. (3.64), an $E°$ value of 0.574 V for the $PoCl_6^{2-}$/Po(s) oxidation couple has been calculated.

$$PoCl_6^{2-} + 4e^- \rightleftharpoons Po(s) + 6Cl^- \quad (3.64)$$

This $E°$ value is in very good agreement with the values (corrected to zero ionic strength) of 0.57 V determined from the data of Bagnall and Freeman (1956a) in 1 mol L^{-1} HCl at 22°C and 0.591 and 0.571 V determined from the measurements of Eichelberger et al. (1965) in 1.0 and 1.5 mol L^{-1} HCl, respectively, at 25°C.

3.2.16 $PoOHCl_4^-$

Preliminary absorption spectroscopy studies of polonium complex ion formation in hydrochloric acid were carried out by McCluggage (1949). The equilibrium between chloride complexes of polonium, responsible for absorption maxima at 344 and 418 μm, were then investigated by Hunt (1954), using an adaptation of the theoretical approach suggested by Bent and French (1941). These studies have been summarized by Moyer (1956) and relate to ambient temperature. The theoretical background behind the studies is contained in Appendix 2.

At high concentrations of hydrochloric acid (>2 mol L^{-1}), there was no appreciable change in absorbance at 418 μm. The molar absorptivity at this wavelength in the more concentrated acid solutions, agreed, within experimental limits, with the value found in 1.2 mol L^{-1} solution. On the basis of other work (Staritzky, 1951; Bagnall et al., 1955; Bagnall and Freeman, 1956a; Eichelberger et al., 1965), the polonium species existing in such solutions is $PoCl_6^{2-}$.

In dilute acid, the absorbances at both 344 and 418 μm were dependent on the chloride and proton concentrations used in the experiments. The effect of reducing the proton concentration in 0.244 mol L^{-1} chloride solutions caused a lowering in the absorbance at 418 μm and an associated increase in the absorbance at 344 μm. Thus a decrease in the proton concentration favors the formation of a complex that absorbs at 344 μm. The data can be interpreted according to Eq. (3.65).

$$Po(OH)_a Cl_{(6-n)}^{(n-a-2)} + nCl^- + aH^+ \rightleftharpoons PoCl_6^{2-} + aH_2O \qquad (3.65)$$

From experiments performed in acid concentrations greater than 2 mol L^{-1}, the molar absorptivity for $PoCl_6^{2-}$ was found to be 1.058×10^4.

The effect of varying the proton concentration on the absorbances at 344 and 418 μm is shown in Table 3.6. For the data given in the table, the chloride concentration was maintained at 0.244 mol L^{-1}. Additionally, the $PoCl_6^{2-}$ concentration was determined from the absorbance at 418 μm and its molar absorptivity. Similarly, the concentration of $Po(OH)_a Cl_{(6-n)}^{(n-a-2)+}$ was determined from the difference between the total polonium and $PoCl_6^{2-}$ concentrations, with the molar absorptivity for $Po(OH)_a Cl_{(6-n)}^{(n-a-2)+}$ calculated to be 4865.

From Eq. (3.65), it follows that

$$\log C_1 + n\log\left[Cl^-\right] = \log\frac{A_{418}}{A_{344}} - a\log\left[H^+\right] \qquad (3.66)$$

where $C_1 = K\,\varepsilon_{418}/\varepsilon_{344}$ (see Appendix 2). At a constant chloride concentration, a plot of log $[H^+]$ against log (A_{418}/A_{344}) should give a straight

Table 3.6 Absorbances at 344 and 418 μm of polonium chloride complexes with varying concentration of protons.

		Absorbance			
$[H^+]$ mol L^{-1}	[Po] mmol L^{-1}	A_{418}	A_{344}	$[PoCl_6^{2-}]$ mmol L^{-1}	$\left[Po(OH)_a Cl_{(6-n)}^{(n-a-2)+}\right]$ mmol L^{-1}
0.244	0.2225	1.570	0.347	0.1484	0.0741
0.195	0.0383	0.262	0.075	0.0248	0.0135
0.146	0.0383	0.223	0.086	0.0211	0.0172
0.098	0.0385	0.163	0.098	0.0154	0.0231

line with the slope equal to *a*, as is shown in Fig. 3.21. The line of best fit in Fig. 3.21 has an intercept of -1.32 ± 0.01 and a slope of 1.09 ± 0.01 [$r^2 = 0.9988$; $P \leq 0.001$; $n = 4$]. From this it can be deduced that the stoichiometric coefficient (*a*) for protons in Eq. (3.65) is 1.

Additional data for experiments at varying concentrations of protons and chloride are summarized in Table 3.7. From these data and Eq. (3.66), a plot of log [Cl^-] against log (A_{418}/A_{344}) − log [H^+] should

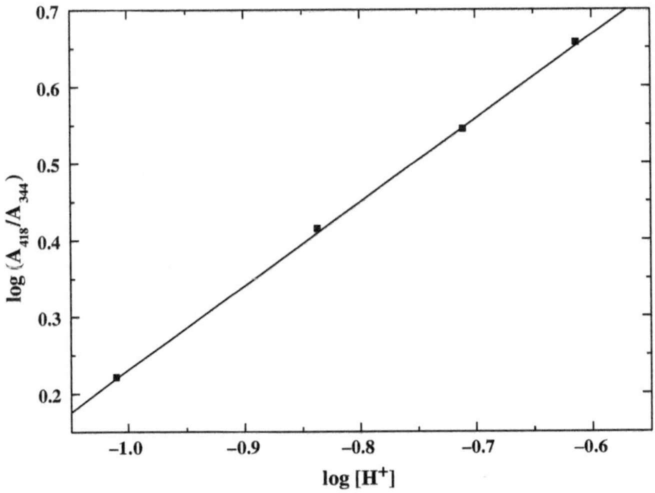

Figure 3.21 Plot of log [H^+] versus log (A_{418}/A_{344}) at constant chloride concentration.

Table 3.7 Absorbances of polonium chloride complexes at 344 and 418 μm with varying concentration of protons.

			Absorbance			
[H^+] M	[Cl^-] M	[Po] mM	A_{418}	A_{344}	[$PoCl_6^{2-}$] mM	[$Po(OH)_aCl_{(6-n)}^{(n-a-2)+}$] mM
0.488	0.488	0.2250	2.207	0.021	0.2086	0.0164
0.366	0.366	2.234	2.070	0.112	0.1957	0.0278
0.244	0.244	0.2225	1.570	0.347	0.1484	0.0741
0.304	0.304	0.1113	0.959	0.117	0.0906	0.0207
0.183	0.183	0.1107	0.466	0.247	0.0440	0.0667
0.152	0.152	0.0553	0.137	0.129	0.0130	0.0424
0.195	0.244	0.0383	0.262	0.075	0.0248	0.0135
0.146	0.244	0.0383	0.223	0.086	0.0211	0.0172
0.098	0.244	0.0385	0.163	0.098	0.0154	0.0231

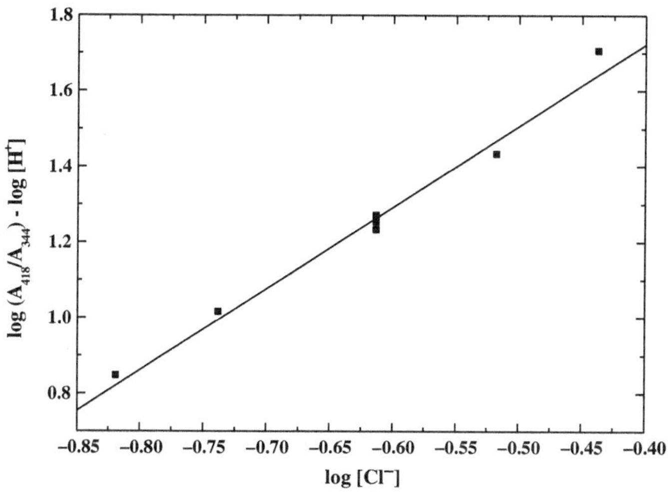

Figure 3.22 Plot of log [Cl$^-$] versus log (A_{418}/A_{344}) − log [H$^+$] at varying proton and chloride concentrations.

give a straight line with slope equal to *n*, as shown in Fig. 3.22. Moyer (1956) omitted points 1 and 2 because of the small absorbance observed at 344 μm for the points and the uncertainty in the correction for the effect of free chlorine. In this work, only point 1 has been omitted since the absorbance at 344 μm for the 0.366 mol L^{-1} solution is considered to be significant. The line of best fit in Fig. 3.22 has an intercept of 2.59 ± 0.07 and a slope of 2.2 ± 0.1 [$r^2 = 0.9837$; $P \leq 0.001$; $n = 8$], which corresponds to a stoichiometric coefficient (*n*) for chloride in Eq. (3.65) of 2.

Substitution of the values determined for *a* and *n* into Eq. (3.65) gives:

$$PoOHCl_4^- + 2Cl^- + H^+ \rightleftharpoons PoCl_6^{2-} + H_2O \quad (3.67)$$

Additionally, the intercept from Fig. 3.22 is equivalent to [log $K°$ − log ($\varepsilon_{344}/\varepsilon_{418}$)]. The value of log $K°$ derived for Eq. (3.67) from this expression and the values of ε_{344} and ε_{418} given above is:

$$\log K°(PoOHCl_4^-) = 2.25 \pm 0.07$$

Use of this log $K°$ and the $\Delta G_f°$ values for $PoCl_6^{2-}$, H_2O and Cl^- gives a Gibbs energy for $PoOHCl_4^-$ of:

$$\Delta G_f°(PoOHCl_4^-) = -(527.5 \pm 0.7)\ \text{kJ mol}^{-1}$$

3.2.17 $PoOCl_4^{2-}$ and $PoO(OH)Cl_2^-$

Suganuma and Hataye (1981) studied the hydrolysis of polonium(IV) in chloride solutions, using solvent extraction with dithizone (H_2Dz) as the extractant at ambient temperature. They found that plots of log $[Cl^-]$ against log D (at constant $[H^+]$ and $[H_2Dz]$), where D is the distribution ratio, gave a slope of -3.5. Similarly, plots of log $[H_2Dz]_{org}$ against log D (at constant $[Cl^-]$ and $[H^+]$) and a plot of pH against log $D - 2\log [H_2Dz]_{org}$ (at constant $[Cl^-]$) both gave slopes of two. This behavior is consistent with the following reactions occurring in the solutions as the pH increases ($0 < pH < 4$).

$$PoCl_6^{2-} + 2H_2Dz_{org} \rightleftharpoons PoCl_2(HDz)_{2\ org} + 2H^+ + 4Cl^- \tag{3.68}$$

$$PoOHCl_4^- + 2H_2Dz_{org} \rightleftharpoons PoOHCl(HDz)_{2\ org} + 2H^+ + 3Cl^- \tag{3.69}$$

$$PoOCl_4^{2-} + 2H_2Dz_{org} \rightleftharpoons PoO(HDz)_{2\ org} + 2H^+ + 4Cl^- \tag{3.70}$$

As the pH is further increased ($4 < pH < 5$), the solvent extraction data are consistent with the following two reactions:

$$PoOCl_4^{2-} + 2H_2O \rightleftharpoons H_2PoO_3(aq) + 2H^+ + 4Cl^- \tag{3.71}$$

$$PoO(OH)Cl_2^- + H_2O \rightleftharpoons H_2PoO_3(aq) + H^+ + 2Cl^- \tag{3.72}$$

Suganuma and Hataye (1981) determined stability constants (log K) for Eqs. (3.69) and (3.70) of -8.7 and -4.6, respectively, in 1.0 M (H, Na)Cl. Use of the Davies equation (Davies, 1962) gives respective log $K°$ values of -8.9 and -4.8. From these log $K°$ values and the $\Delta G_f°$ values for H_2PoO_3, Cl^- and H_2O, the calculated $\Delta G_f°$ values for $PoOCl_4^{2-}$ and $PoO(OH)Cl_2^-$ are:

$$\Delta G_f°(PoOCl_4^{2-}) = -(493.8 \pm 6.3)\ kJ\ mol^{-1}$$

$$\Delta G_f°(PoO(OH)Cl_2^-) = -(445.1 \pm 6.3)\ kJ\ mol^{-1}$$

The stability constant determined for Eq. (3.71)

$$PoO^{2+} + 4Cl^- \rightleftharpoons PoOCl_4^{2-} \tag{3.73}$$

from the derived Gibbs energy values for $PoOCl_4^{2-}$, PoO^{2+} and Cl^- is:

$$\log K^\circ(PoOCl_4^{2-}) = 6.4 \pm 0.3$$

where the uncertainty has been estimated in this review. This constant is less than that proposed by Starik and coworkers [log $K = 8.53$ in $1\ mol\ L^{-1}$ $H(Cl,\ ClO_4)$ (Starik et al., 1964b) and log $K = 8.85$ in $4{-}6\ mol\ L^{-1}$ $H(Cl,\ ClO_4)$ (Starik and Ampelogova, 1965)]. The species proposed by these latter authors was the neutral species $PoCl_4(aq)$ rather than the anionic species $PoOCl_4^{2-}$ as proposed by Suganuma and Hataye (1981). The average stepwise stability constant relative to Eq. (3.73) is $6.4/4 = 1.6$ (i.e., the overall stability constant divided by the number of chloride ions in the species), whereas that for Eq. (3.61) is $14.55/6 = 2.43$. It is usual for the average stepwise stability constant to decrease as the ligand number (in this case, chloride ions) increases. When the opposite occurs, it is almost always associated with a change in structure. In the case of the polonium(IV) chloride complexes, this change in structure is the loss of the oxygen atom when transitioning from the $PoOCl_4^{2-}$ species to the $PoCl_6^{2-}$ species (the cation changes from divalent to tetravalent). This structural change was not apparent in the stability constant data proposed by Starik and coworkers and, as such, their data are not accepted.

3.2.18 $PoBr_2(s)$

Zhdanov (1985) gives a value of $155\ J\ mol^{-1}\ K^{-1}$ for the S_f° of $PoBr_2(s)$ from which a ΔS_f° value of $-60.0\ J\ mol^{-1}\ K^{-1}$ is calculated using Eq. (3.18). From this ΔS_f° and the line of best fit given in Fig. 3.13, the calculated value for the Gibbs energy of $PoBr_2(s)$ is:

$$\Delta G_f^\circ(PoBr_2(s)) = -(52.2 \pm 5.9)\ kJ\ mol^{-1}$$

An uncertainty of $2\ J\ mol^{-1}\ K^{-1}$ was assigned to the entropy of Zhdanov (1985) by the present review.

3.2.19 $PoBr_4(s)$

Zhdanov (1985) quotes a ΔH_f° value of $-190.4\ kJ\ mol^{-1}$ for $TeBr_4(s)$. Use of this value and the regression equation from Fig. 3.15 gives a ΔH_f° of $-199.8\ kJ\ mol^{-1}$ for $PoBr_4(s)$. Zhdanov (1985) also quotes an S_f° of $230\ J\ mol^{-1}\ K^{-1}$ for $PoBr_4(s)$. Use of the ΔS_f° calculated from this value (Eq. (3.18)) and the ΔH_f° for $PoBr_4(s)$ gives a ΔG_f° of $-158.8\ kJ\ mol^{-1}$

for $PoBr_4$(s) (Eq. (3.6)). Similarly, a ΔG_f° of -148.5 kJ mol^{-1} for $PoBr_4$(s) is calculated using the ΔH_f° for $PoBr_4$(s) and the regression equation from Fig. 3.5. The average of the two ΔG_f° values is accepted for the Gibbs energy of $PoBr_4$(s).

$$\Delta G_f^\circ(\mathrm{PoBr_4(s)}) = -(153.7 \pm 5.2)\ \mathrm{kJ\ mol^{-1}}.$$

The uncertainty is assigned to span the range in the two values.

3.2.20 PoI_2(s) and PoI_4(s)

Hisham and Benson (1992) derived an empirical relationship between the enthalpies of formation of solid halides and the corresponding gas-phase halide ions. The authors demonstrated that values of $\Delta(MX_n)$ for any two metals with the same valence were linearly related, where

$$\Delta(\mathrm{MX_n}) = \Delta H_f^\circ(\mathrm{MX_n(s)}) - \Delta H_f^\circ(\mathrm{X^-(g)}) \tag{3.74}$$

The corresponding ΔH_f° values for $PoCl_2$(s) and $PoBr_2$(s) calculated using Eq. (3.6) and the ΔG_f° and ΔS_f° values above, are -177.4 and -68.3 kJ mol^{-1}, respectively. Use of these ΔH_f° values and -233.3, -220.3, and -197.3 kJ mol^{-1} for the ΔH_f° of Cl^-(g), Br^-(g) and I^-(g) (Hisham and Benson, 1992), respectively, gives a ΔH_f° value of 128.1 kJ mol^{-1} for PoI_2(s). In turn, use of this value and the line of best fit given in Fig. 3.5 leads to a Gibbs energy for PoI_2(s) of:

$$\Delta G_f^\circ(\mathrm{PoI_2(s)}) = (95.2 \pm 4.5)\ \mathrm{kJ\ mol^{-1}}$$

An uncertainty of 5 kJ mol^{-1} was assigned to the enthalpy of PoI_2(s) by the present review.

An average value of -41.1 kJ mol^{-1} for the ΔH_f° of TeI_4(s) has been determined from the data given in Haag et al. (1979). Use of this value and the regression equation from Fig. 3.15 gives a value of -53.4 kJ mol^{-1} for the ΔH_f° of PoI_4(s). In turn, the calculated ΔG_f° value for PoI_4(s) using this ΔH_f° and the regression equation from Fig. 3.5 is -39.7 kJ mol^{-1}.

Borzhim et al. (1979) determined a ΔG_f° for TeI_4(s) of -43.4 kJ mol^{-1} at 20°C. Use of this value and the regression equation from Fig. 3.14 gives a ΔG_f° of -53.6 kJ mol^{-1} for PoI_4(s). The average of these two ΔG_f° values is selected:

$$\Delta G_f^\circ(\mathrm{PoI_4(s)}) = -(46.7 \pm 7.0)\ \mathrm{kJ\ mol^{-1}}$$

The uncertainty is assigned to span the range in the two values.

3.2.21 PoI_5^- and PoI_6^{2-}

Bagnall et al. (1956) investigated the solubility of PoI_4(s) in hydriodic acid and determined a solubility constant (log $K°$) of −2.23 for Eq. (2.5) at 22°C. Use of this log $K°$ and the $\Delta G_f°$ values for PoI_4(s) and I^- gives a $\Delta G_f°$ for PoI_6^{2-} of −137.4 kJ mol^{-1}.

Bigelis et al. (1978) measured an $E°$ value for the TeI_6^{2-}/Te couple in hydriodic acid. From the value obtained (0.415 V) and the $\Delta G_f°$ value for I^-, the calculated $\Delta G_f°$ for TeI_6^{2-} is −150.2 kJ mol^{-1}. Use of this value and the regression equation from Fig. 3.14 gives a $\Delta G_f°$ of −156.9 kJ mol^{-1} for PoI_6^{2-}. Although this latter value is not inconsistent with that obtained from the data of Bagnall et al. (1956), the former value is preferred because it has been derived from direct measurements of PoI_6^{2-}. On the basis of the measured potential values of Bigelis et al. (1978), it is likely that their $E°$ value should more likely be in the range of 0.46–0.47 V.

Therefore the selected Gibbs energy of PoI_6^{2-} is:

$$\Delta G_f°\left(PoI_6^{2-}\right) = -(137.4 \pm 7.3)\ \text{kJ mol}^{-1}$$

The uncertainty has been determined on the basis of a 2 kJ mol^{-1} uncertainty being assigned by the present review to the Gibbs energy of the solubility reaction of Bagnall et al. (1956).

From measurements using low acid concentrations ([HI] < 0.02 M), Bagnall et al. (1956) found a solubility constant (log $K°$) of −4.17 for Eq. (2.6) at 22°C. Use of this log $K°$ and the $\Delta G_f°$ values for PoI_4(s) and I^- leads to a Gibbs energy value for PoI_5^- of:

$$\Delta G_f°\left(PoI_5^-\right) = -(74.6 \pm 7.3)\ \text{kJ mol}^{-1}$$

Again, the uncertainty has been assigned by the present review in the same manner as for PoI_6^{2-}.

3.2.22 PoS(s)

Bagnall and Robertson (1957) found the black precipitate that formed when H_2S(g) was passed through aqueous solutions of polonium dichloride or tetrachloride was PoS(s). They used a variety of analytical techniques to determine the Po:S ratio since X-ray powder photographs were too poor to index.

The solubility product was determined by precipitating the compound from solutions of varying HCl concentration, which had been saturated with $H_2S(g)$ at ambient temperature. The sulfide ion concentration was calculated from the solubility data of Kendall and Andrews (1921) and the known dissociation constants of $H_2S(g)$. Activity corrections were not applied. The log $K_{sp}°$ obtained was −28.26.

The equilibrium sulfide concentrations present in the solutions of Bagnall and Robertson (1957) were recalculated using more recent data; 3.38 g L^{-1} for the solubility of $H_2S(g)$ in water at 25°C (Aylward and Findlay, 1974) and the thermochemical data for sulfur species contained in Appendix 1. A solubility constant of:

$$\log K_{sp}°(\mathrm{PoS(s)}, (3.72)) = -27.26$$

was derived using these values and the Po^{2+} and H^+ concentrations given by Bagnall and Robertson (1957).

For Eq. (3.75),

$$\mathrm{PoS(s)} \rightleftharpoons \mathrm{Po}^{2+} + \mathrm{S}^{2-} \tag{3.75}$$

a Gibbs energy value for PoS(s) was calculated using the derived log $K_{sp}°$ and the $\Delta G_f°$ values for Po^{2+} and S^{2-}.

$$\Delta G_f°(\mathrm{PoS(s)}) = (61.8 \pm 8.7)\ \mathrm{kJ\ mol^{-1}}$$

An uncertainty of 2 kJ mol^{-1} was assigned to the Gibbs energy of the solubility reaction by the present review.

3.2.23 $PoSO_4(s)$

Zikovsky (1998) precipitated polonium from solution using $BaSO_4(s)$ as the carrier. The solubility of $PoSO_4(s)$ was calculated (assuming that an isomorphous precipitate was formed) using Eq. (3.76) (ambient temperature), where K and K^* are the solubility products of $BaSO_4(s)$ and $PoSO_4(s)$, respectively, and D is the distribution coefficient, as described by Driessens (1984).

$$K^* = D \cdot K \tag{3.76}$$

The reported value of 10 for the K_{sp} of $BaSO_4(s)$, however, is clearly in error. A review of the literature (Brown et al., 2015) suggests that this value is in fact the log K of reaction (3.77).

$$BaSO_4(s) \rightleftharpoons Ba^{2+} + SO_4^{2-} \quad (3.77)$$

The value for the K of $PoSO_4(s)$, obtained using the correct value of 8.193×10^9 for the K of $BaSO_4(s)$, is 7.943×10^8. This is equivalent to a solubility product of:

$$\log K_{sp}°(PoSO_4(s)) = -8.90$$

for $PoSO_4(s)$ in Eq. (3.78).

$$PoSO_4(s) \rightleftharpoons Po^{2+} + SO_4^{2-} \quad (3.78)$$

Use of this log $K_{sp}°$ and the $\Delta G_f°$ values for Po^{2+} and SO_4^{2-} gives:

$$\Delta G_f°(PoSO_4(s)) = -(670.7 \pm 8.6)\ kJ\ mol^{-1}$$

for $PoSO_4(s)$, with respect to the experimental conditions used by Zikovsky (1998). Again, an uncertainty of 2 kJ mol^{-1} was assigned to the Gibbs energy of the solubility reaction by the present review.

Zikovsky also studied the precipitation behavior and solubilities of polonium(II) arsenites, chromates, iodides, molybdates, sulfites, and vanadates. Unlike sulfate, however, these experiments were not carried out in acid solution. It is therefore unlikely, based on the conclusions of this current review, that the polonium was present as Po^{2+} in these experiments.

3.2.24 $PoOSO_4$(aq) and $PoO(SO_4)_2^{2-}$

Ampelogova (1973) studied the complexation of polonium(IV) by sulfate using solvent extraction at ambient temperature. Stability constants (β_n) of 29 and 2500 were found for Eqs. (3.79) and (3.80), respectively, in 2 mol L^{-1} H(ClO_4, HSO_4).

$$PoO^{2+} + SO_4^{2-} \rightleftharpoons PoOSO_4(aq) \quad (3.79)$$

$$PoO^{2+} + 2SO_4^{2-} \rightleftharpoons PoO(SO_4)_2^{2-} \quad (3.80)$$

The stability constant of $PoOSO_4$(aq) was corrected to zero ionic strength, based on the difference between data for other divalent metal ions in perchlorate media (at the same ionic strength) and those at zero ionic strength (Sillén and Martell, 1964, 1971; Högfeldt, 1982). The stability constant for $PoO(SO_4)_2^{2-}$ leads to a second stepwise constant that is larger than the first constant, which appears unlikely, particularly in attaining consistency with respect to the measured solubilities of

$Po(SO_4)_2 \cdot H_2O(s)$, and $(PoO)_2OSO_4(s)$. Thus the stability constant for $PoO(SO_4)_2^{2-}$ is based on the second stepwise constants for other divalent metals [e.g., Pb^{2+}, Cd^{2+}, Zn^{2+} (Powell et al., 2009, 2011, 2013)]. This leads to log β_n° values of:

$$\log \beta_1°(PoOSO_4(aq)) = 2.56 \pm 0.30$$

$$\log \beta_2°\left(PoO(SO_4)_2^{2-}\right) = 3.61 \pm 0.30$$

for $PoOSO_4$ and $PoO(SO_4)_2^{2-}$, respectively, in Eqs. (3.79) and (3.80). The former value is not inconsistent with the value determined by Katzlberger (2000), who determined a stability constant for a reaction similar to Eq. (3.79), but with HSO_4^- instead of SO_4^{2-}, of log $K = 3.65$ in 1 mol L^{-1} H_2SO_4–$HClO_4$ solutions. This constant would appear consistent with that derived in the present study when considering the differences in ionic strength and the use of SO_4^{2-} or HSO_4^- in the chemical reaction. On the basis of these corrected log β_n° values and the ΔG_f° values for PoO^{2+} and SO_4^{2-}, the calculated ΔG_f° values are:

$$\Delta G_f°(PoOSO_4(aq)) = -(691.1 \pm 5.4)\ \text{kJ mol}^{-1}$$

$$\Delta G_f°\left(PoO(SO_4)_2^{2-}\right) = -(1441.1 \pm 5.5)\ \text{kJ mol}^{-1}$$

The uncertainties for the stability constants were assigned by the present review.

3.2.25 $Po(SO_4)_2 \cdot H_2O(s)$ and $PoO(SO_4)_3^{4-}$

Bagnall and Freeman (1956b) studied the solubility of polonium disulfate in various concentrations of sulfuric acid at ambient temperature. From their measurements, a plot of the logarithm of the acid concentration versus the logarithm of the solubility (Fig. 3.23) shows that there is a change in solution speciation at point B.

The line of best fit for the A-B portion of the graph has a slope of 0.4324 ± 0.0006 [$r^2 = 1$; $P \leq 0.001$; $n = 3$], and therefore two moles of polonium will be dissolved by one mole of sulfate. Eq. (3.81) describes the reaction.

$$2Po(SO_4)_2 \cdot H_2O(s) + HSO_4^- \rightleftharpoons PoO(SO_4)_2^{2-} + PoO(SO_4)_3^{4-} + 5H^+ \quad (3.81)$$

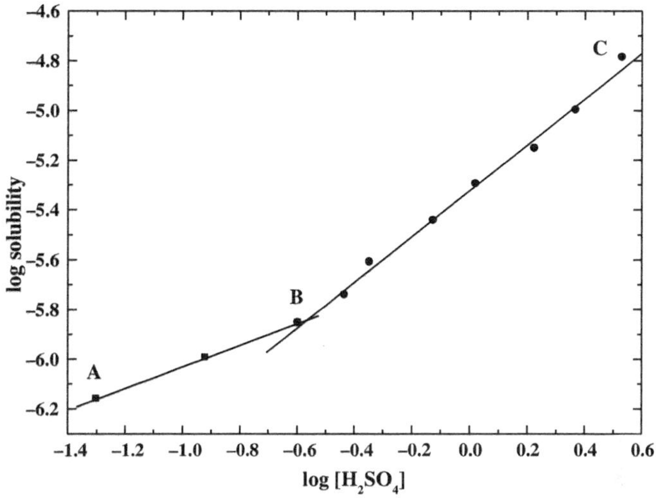

Figure 3.23 Plot of log $[H_2SO_4]$ versus log solubility for $Po(SO_4)_2 \cdot H_2O(s)$.

The line of best fit for the B-C portion of the graph has a slope of 0.92 ± 0.03 [$r^2 = 0.9943$; $P \leq 0.001$; $n = 8$], and therefore the reaction can be written as follows:

$$Po(SO_4)_2 \cdot H_2O(s) + HSO_4^- \rightleftharpoons PoO(SO_4)_3^{4-} + 3H^+ \quad (3.82)$$

since one mole of polonium will be dissolved by one mole of sulfate.

The line of best fit for the B–C portion of the graph can be used to determine the $PoO(SO_4)_3^{4-}$ concentrations for the three points on the A–B portion. These, in turn, can be used to derive the concentration of $PoO(SO_4)_2^{2-}$ since, in the A–B region,

$$[Po]_T = \left[PoO(SO_4)_2^{2-}\right] + \left[PoO(SO_4)_3^{4-}\right] \quad (3.83)$$

The stability constant for the stepwise Eq. (3.84)

$$PoO(SO_4)_2^{2-} + SO_4^{2-} \rightleftharpoons PoO(SO_4)_3^{4-} \quad (3.84)$$

can be written as

$$K_3{}^\circ = \frac{[PoO(SO_4)_3^{4-}]}{[PoO(SO_4)_2^{2-}]\ [SO_4^{2-}]} \quad (3.85)$$

and an average $K_3°$ value of 5.62 has been obtained from the known concentration of SO_4^{2-} and the derived concentrations of $PoO(SO_4)_2^{2-}$ and $PoO(SO_4)_3^{4-}$. This is equivalent to:

$$\log K_3°\left(PoO(SO_4)_3^{4-}\right) = 0.75 \pm 0.24$$

for Eq. (3.84) using the specific ion interaction theory (Grenthe et al., 1992). Considering the stepwise stability constants of the sulfate complexes of PoO^{2+}, it is possible that the third complex actually relates to the formation of $Po(SO_4)_3^{2-}$ rather than $PoO(SO_4)_3^{4-}$, but the latter will be used in the present review. Use of this log $K_3°$ and the $\Delta G_f°$ values for $PoO(SO_4)_2^{2-}$ and SO_4^{2-} gives a Gibbs energy value for $PoO(SO_4)_3^{4-}$ of:

$$\Delta G_f°\left(PoO(SO_4)_3^{4-}\right) = -(2189.3 \pm 5.7)\ \text{kJ mol}^{-1}$$

A solubility constant of:

$$\log K_{sp}°\left(Po(SO_4)_2 \cdot H_2O(s)\right) = -9.44 \pm 0.06$$

for Eq. (3.82) has been derived using the specific ion interaction theory (Grenthe et al., 1992) (Fig. 3.24) [$r^2 = 0.9725$; $P \leq 0.001$; $n = 6$]. Use of this log $K_{sp}°$ and the $\Delta G_f°$ values for $PoO(SO_4)_3^{4-}$ and HSO_4^-, gives a Gibbs energy for $Po(SO_4)_2 \cdot H_2O(s)$ of:

$$\Delta G_f°\left(Po(SO_4)_2 \cdot H_2O(s)\right) = -(1487.9 \pm 5.8)\ \text{kJ mol}^{-1}$$

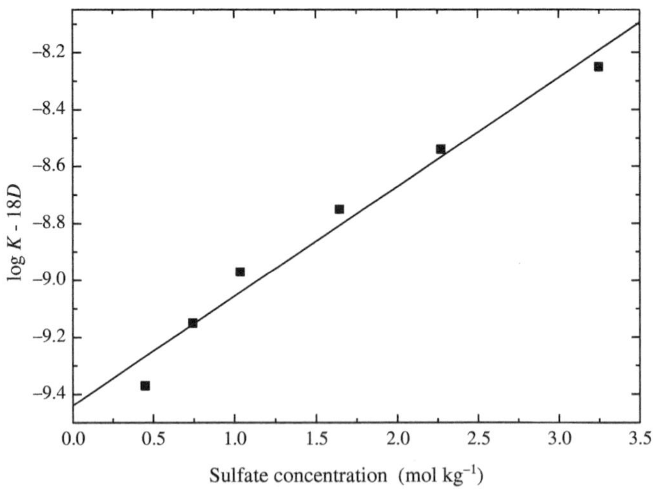

Figure 3.24 Plot of ionic strength versus log $K - 18D$ for $Po(SO_4)_2.H_2O(s)$.

3.2.26 $(PoO)_2OSO_4(s)$

Bagnall and Freeman (1956b) studied the solubility of the basic sulfate $2PoO_2 \cdot SO_3(s)$ $\{(PoO)_2OSO_4(s)$ [equally, the more probable stoichiometry for the phase is $(PoO)_2(OH)_2SO_4(s)]\}$ in various concentrations of sulfuric acid at ambient temperature. From their measurements, a plot of the logarithm of the acid concentration versus the logarithm of the solubility has a slope of 0.55 ± 0.03 [$r^2 = 0.9890$; $P \le 0.001$; $n = 7$] (Fig. 3.25). This indicates that one mole of acid will solubilize two moles of polonium. The equation can therefore be written as follows:

$$(PoO)_2OSO_4(s) + HSO_4^- + H^+ \leftrightarrow 2PoOSO_4(aq) + H_2O \qquad (3.86)$$

A solubility constant of:

$$\log K°\left((PoO)_2OSO_4(s)\right) = -7.75 \pm 0.08$$

has been derived for Eq. (3.86) using the specific ion interaction theory (Grenthe et al., 1992) (Fig. 3.26). A similar calculation (figure not shown) with respect to Eq. (3.87):

$$(PoO)_2OSO_4(s) + 3HSO_4^- \rightleftharpoons 2PoO(SO_4)_2^{2-} + H_2O + H^+ \qquad (3.87)$$

leads to a solubility constant for Eq. (3.86) of:

$$\log K°\left((PoO)_2OSO_4(s)\right) = -9.38 \pm 0.06$$

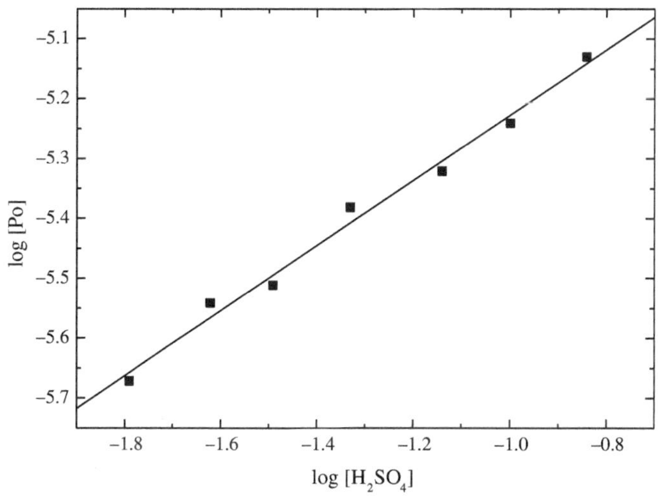

Figure 3.25 Plot of log $[H_2SO_4]$ versus log solubility for $(PoO)_2OSO_4(s)$.

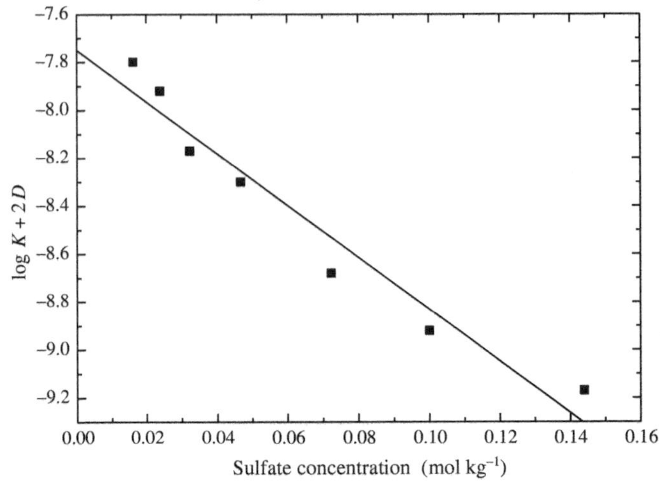

Figure 3.26 Plot of sulfate concentration versus log *K* for $(PoO)_2OSO_4$(s).

The first solubility constant leads to a Gibbs energy of -908 ± 11 kJ mol^{-1}, whereas the second gives -907 ± 12 kJ mol^{-1}. Use of the derived log K° values and the ΔG_f° values for $PoOSO_4$(aq), $PoO(SO_4)_2^{2-}$, HSO_4^-, and H_2O leads to the weighted average value of:

$$\Delta G_f^\circ((PoO)_2OSO_4(s)) = -(907.6 \pm 8.0)\ \text{kJ mol}^{-1}$$

for $(PoO)_2OSO_4$(s). In turn, a solubility constant of:

$$\log K_{sp}^\circ((PoO)_2OSO_4(s)) = -38.8 \pm 0.1$$

has been derived for reaction (3.88) using the ΔG_f° value and those for PoO^{2+}, SO_4^{2-}, OH^-, and H_2O.

$$(PoO)_2OSO_4(s) + H_2O \rightleftharpoons 2PoO^{2+} + 2OH^- + SO_4^{2-} \qquad (3.88)$$

3.2.27 $PoSO_4$(aq)

Joliot (1930) measured the cathodic deposition of divalent polonium in 0.25 mol L^{-1} H_2SO_4 and determined an *E* of 0.63 V for Eq. (3.89) at ambient temperature.

$$PoSO_4(aq) + 2e^- \rightleftharpoons Po(s) + SO_4^{2-} \qquad (3.89)$$

Correction of this E° to zero ionic strength using the Davies equation (Davies, 1962) results in an E° of 0.61 V. From this corrected E° and the

ΔG_f° values for Po(s) and SO_4^{2-}, the Gibbs energy calculated for $PoSO_4(aq)$ is:

$$\Delta G_f^\circ(PoSO_4(aq)) = -(626.3 \pm 3.9)\ \text{kJ mol}^{-1}$$

An uncertainty of 0.02 V was assigned to the zero ionic strength potential by the present review.

Use of this ΔG_f° and the ΔG_f° value for Po^{2+} gives:

$$\log K^\circ(PoSO_4(aq)) = 1.13$$

for Eq. (3.90).

$$Po^{2+} + SO_4^{2-} \rightleftharpoons PoSO_4(aq) \quad (3.90)$$

Thus the complexation of Po^{2+} by sulfate is weak. For example, in 0.25 mol L^{-1} H_2SO_4, $PoSO_4(aq)$ accounts for only 41% of the total polonium. Joliot (1930) also measured the cathodic deposition of polonium(II) in 0.05 mol L^{-1} H_2SO_4, and obtained an electrode potential of 0.65 V at ambient temperature, which is similar to that obtained for the $Po^{2+}/Po(s)$ couple (0.643 V). This result is not surprising since in 0.05 mol L^{-1} H_2SO_4, polonium sulfate accounts for only 12% of the total polonium.

3.2.28 $(PoO)_2OSeO_4(s)$

Bagnall and Freeman (1956b) studied the solubility of the basic selenate $2PoO_2 \cdot SeO_3(s)$ {$(PoO)_2OSeO_4(s)$ [again, this phase might be better expressed as $(PoO)_2(OH)_2SeO_4(s)$]} in various concentrations of selenic acid at ambient temperature. From their measurements, a plot of the logarithm of the acid concentration against the logarithm of the solubility (Fig. 3.27) has a slope of 0.61 ± 0.02 [$r^2 = 0.9962$; $P \leq 0.001$; $n = 7$] at low selenic acid concentrations (≤ 0.56 mol L^{-1}) and a slope of 1.61 ± 0.07 [$r^2 = 0.9943$; $P \leq 0.001$; $n = 5$] at high selenic acid concentrations (≥ 0.8 mol L^{-1}).

At low selenic acid concentrations, therefore, one mole of acid will solubilize two moles of polonium and at high selenic acid concentrations, three moles of acid will solubilize two moles of polonium. The equations can be written as follows:

$$(PoO)_2OSeO_4(s) + HSeO_4^- + H^+ \rightleftharpoons 2PoOSeO_4 + H_2O \quad (3.91)$$

$$(PoO)_2OSeO_4(s) + 3HSeO_4^- \rightleftharpoons 2PoO(SeO_4)_2^{2-} + H_2O + H^+ \quad (3.92)$$

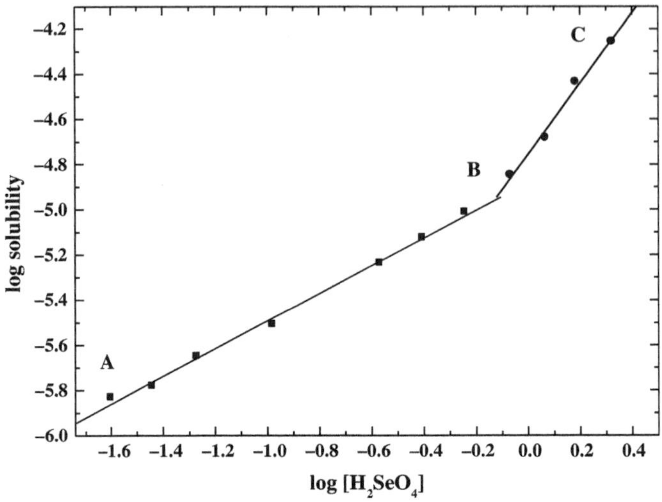

Figure 3.27 Plot of log [H_2SeO_4] versus log solubility for $(PoO)_2OSeO_4(s)$.

A solubility constant for Eq. (3.91) of:

$$\log K^\circ\left((PoO)_2OSeO_4(s)\right) = -8.53 \pm 0.09$$

has been derived using the specific ion interaction theory (Grenthe et al., 1992) (Fig. 3.28) [$r^2 = 0.9050$; $P \leq 0.001$; $n = 7$]. Use of this log K° and the ΔG_f° values for $PoOSeO_4(aq)$, $HSeO_4^-$ and H_2O gives a Gibbs energy of -608 ± 11 kJ mol^{-1} for $(PoO)_2OSeO_4(s)$.

Similarly, a solubility constant relating to Eq. (3.92) of:

$$\log K^\circ\left((PoO)_2OSeO_4(s)\right) = -10.82 \pm 0.11$$

[$r^2 = 0.9392$; $P \leq 0.01$; $n = 5$] (Fig. 3.29) has been derived. Use of this log K° and the ΔG_f° values for $PoO(SeO_4)_2^{2-}$ (see Section 3.2.29), $HSeO_4^-$, and H_2O gives a ΔG_f° of -603 ± 13 kJ mol^{-1} for $(PoO)_2OSeO_4(s)$. The weighted average of these two ΔG_f° values is:

$$\Delta G_f^\circ\left((PoO)_2OSeO_4(s)\right) = -(605.7 \pm 8.6)\ \text{kJ mol}^{-1}$$

A solubility constant of:

$$\log K_{sp}^\circ\left((PoO)_2OSeO_4(s)\right) = -39.2 \pm 0.4$$

has been derived for Eq. (3.93) using the average ΔG_f° and the ΔG_f° values for PoO^{2+}, SeO_4^{2-}, OH^- and H_2O.

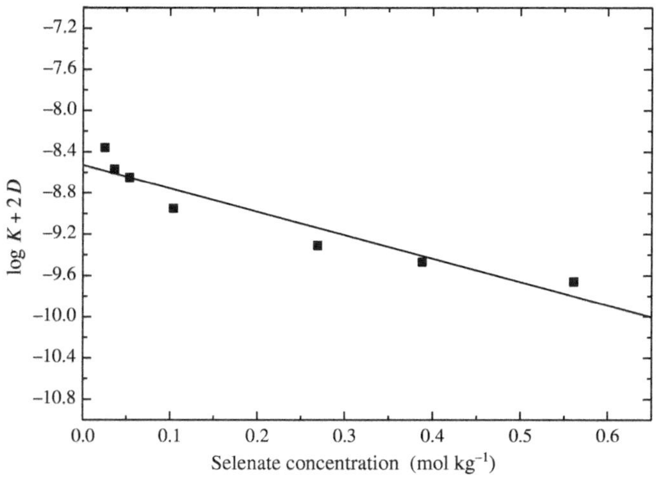

Figure 3.28 Plot of ionic strength versus log $K + 2D$ for $(PoO)_2OSeO_4(s)$.

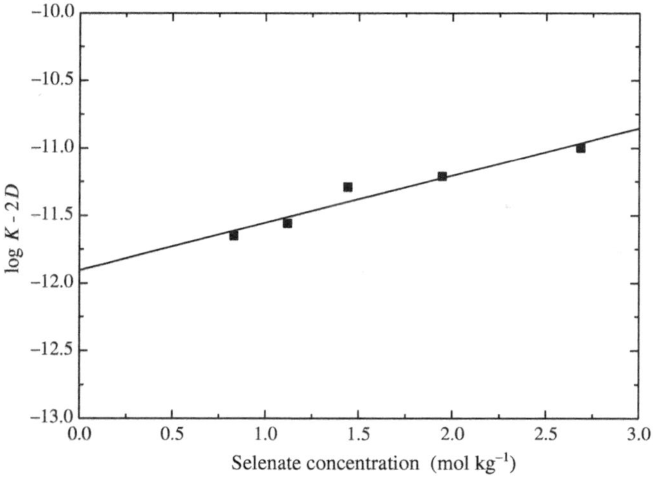

Figure 3.29 Plot of ionic strength versus log $K - 2D$ for $(PoO)_2OSeO_4(s)$.

$$(PoO)_2OSeO_4(s) + H_2O \rightleftharpoons 2PoO^{2+} + 2OH^- + SeO_4^{2-} \quad (3.93)$$

3.2.29 $PoOSeO_4(aq)$ and $PoO(SeO_4)_2^{2-}$

The stability constant (β_1) for a polonium(IV)—selenate complex, as given by Eq. (3.94), was estimated from the equivalent polonium(IV)—sulfate complexes using the unified theory of metal ion complexation (Brown and Sylva, 1987). The second stability constant (β_2) was estimated, relative to Eq. (3.95), from the discontinuity point in the solubility behavior observed by Bagnall and Freeman (1956b). The discontinuity relates to the point where the solution speciation transfers from being dominated by $PoOSeO_4(aq)$ to $PoO(SeO_4)_2^{2-}$, thus allowing determination of the stepwise stability constant for the formation of $PoO(SeO_4)_2^{2-}$ from $PoOSeO_4(aq)$.

$$PoO^{2+} + SeO_4^{2-} \rightleftharpoons PoOSeO_4(aq) \qquad (3.94)$$

$$PoO^{2+} + 2SeO_4^{2-} \rightleftharpoons PoO(SeO_4)_2^{2-} \qquad (3.95)$$

The stability constants ($\log \beta_n°$) estimated are:

$$\log \beta_1°(PoOSeO_4(aq), (3.92)) = 2.43 \pm 0.30$$

$$\log \beta_2°\left(PoO(SeO_4)_2^{2-}, (3.93)\right) = 2.57 \pm 0.42$$

for the formation of $PoOSeO_4(aq)$ and $PoO(SeO_4)_2^{2-}$, respectively, via Eqs. (3.94) and (3.95). The uncertainties for both stepwise constants were assigned a value of 0.3 log units by the present review.

On the basis of these $\log \beta_n°$ values and the $\Delta G_f°$ values for PoO^{2+} and SeO_4^{2-}, the calculated $\Delta G_f°$ values for $PoOSeO_4(aq)$ and $PoO(SeO_4)_2^{2-}$ are:

$$\Delta G_f°(PoOSeO_4(aq)) = -(385.8 \pm 5.6)\ \text{kJ mol}^{-1}$$

$$\Delta G_f°\left(PoO(SeO_4)_2^{2-}\right) = -(826.1 \pm 6.4)\ \text{kJ mol}^{-1}$$

3.2.30 $Ag_2PoO_3(s)$

The suitability of linear free energy relationships to describe correlations between thermochemical data was demonstrated in Sections 3.2.12 and 3.2.14. From a plot of $\log K°$ (HXO_3^-) versus $\log K_{sp}°$ ($Ag_2XO_3(s)$), where X is S, Se, or Te, it is possible to estimate a value for $Ag_2PoO_3(s)$ (Fig. 3.30). The $\log K°$ values for HXO_3^- and the $\log K_{sp}°$ ($Ag_2XO_3(s)$) values for the respective chalcogens are given in Table 3.8.

Fig. 3.30 illustrates that there is excellent agreement between these parameters, giving an intercept of −1.5 (0.4) and a slope of -1.72 ± 0.04 [$r^2 = 0.9994$; $P \leq 0.05$; $n = 3$].

The solubility of $Ag_2PoO_3(s)$ can be described by Eq. (3.96).

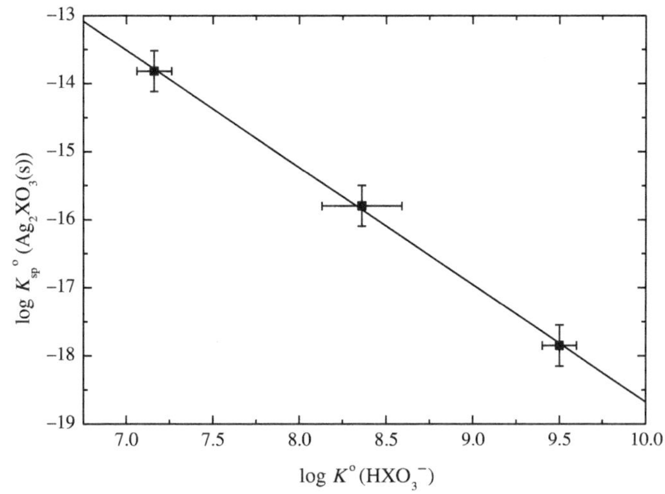

Figure 3.30 Plot of log *K*° (HXO_3^{2-}) (X is S, Se, or Te) versus log K_{sp}° (Ag_2XO_3).

Table 3.8 log *K*° values of HXO_3^- and log K_{sp}° values of Ag_2XO_3 (X is S, Se, and Te) at 25°C and zero ionic strength.

Chalcogen	log *K*° (HXO_3^-)[a]	log K_{sp}° (Ag_2XO_3(s))[b]
S	7.16	− 13.82
Se	8.36	− 15.80
Te	9.50	− 17.85
Po	13.56	

[a]S and Se data from Högfeldt (1982), Te data from Masson (1976), and Po data from this work (see $HPoO_3^{2-}$).
[b]S and Se data from Sillén and Martell (1964) and Te data from Högfeldt (1982).

$$Ag_2PoO_3(s) \rightleftharpoons 2Ag^+ + PoO_3^{2-} \quad (3.96)$$

The solubility constant (Eq. (3.94)) for $Ag_2PoO_3(s)$ determined from the linear fit is:

$$\log K_{sp}°(Ag_2PoO_3(s)) = -24.8$$

Use of the calculated log K_{sp}° and the ΔG_f° values for Ag^+ and PoO_3^{2-} gives:

$$\Delta G_f°(Ag_2PoO_3(s)) = -(231.7 \pm 3.6)\ \text{kJ mol}^{-1}$$

An uncertainty of 0.2 log units was assigned to the solubility constant by the present review.

3.2.31 PbPo(s), HgPo(s), ZnPo(s), NiPo(s), and Ag_2Po(s)

The general equation for the formation of metal polonides is given by Eq. (3.95):

$$MPo(s) \rightleftharpoons M^{2+} + Po^{2-} \quad (3.97)$$

A large number of metal polonides have been prepared and analyzed using X-ray diffraction (Moyer, 1956; Bagnall, 1957; Abakumov, 1982). Of the metals that have been studied, solubility data exist for the sulfides, selenides, and tellurides of Pb, Hg, Zn, Ni, and Ag only. It can be shown that a relationship exists between these solubilities and the dissociation constant of the respective hydrogen chalcogenide (HX^-). This is illustrated in Fig. 3.31 using the data contained in Table 3.9.

The lines of best fit for each metal are summarized in Table 3.10. From these, the following log K_{sp}° values have been obtained for metal polonides described by Eq. (3.97):

$$\log K_{sp}°(PbPo(s), (3.95)) = -51.2 \pm 0.5$$

$$\log K_{sp}°(HgPo(s), (3.95)) = -73.4 \pm 0.5$$

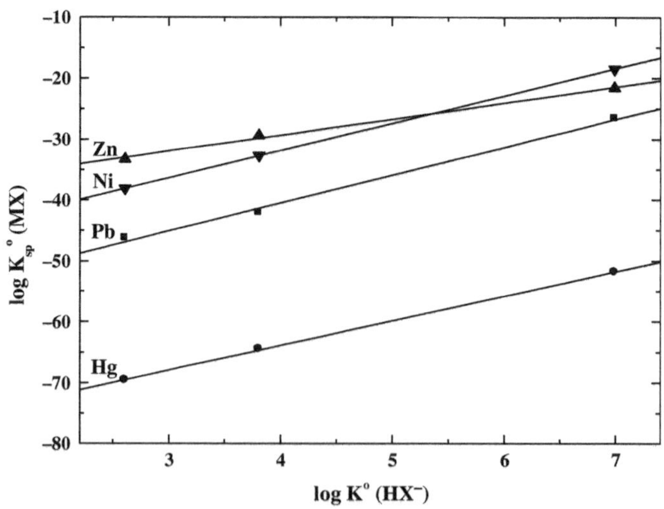

Figure 3.31 Plots of log K° (HX^-) versus log K_{sp}° (MX) for metal chalcogenides.

Table 3.9 Literature data for metal chalcogenide solubilities [Ringbom (1953) (S); Buketov et al. (1964) (Se and Te)].

Chalcogen	log $K°$ (HX^-)	Solubility [log $K_{sp}°$ (MX)]				
		Pb	Hg	Zn	Ni	Ag
S	6.99	−26.6	−51.8	−21.6	−18.5	−49.2
Se	3.85	−42.1	−64.5	−29.4	−32.7	−63.7
Te	2.64	−46.3	−69.6	−33.3	−38.1	−71.7

Table 3.10 Least squares regression data for HX^-/MX plots.

Metal	Intercept	Slope	r^2	P	n
Pb	−59 (2)	−4.6 (0.3)	0.9952	≤0.05	3
Hg	−80.3 (0.2)	−4.08 (0.04)	0.9999	≤0.01	3
Zn	−40.0 (0.8)	−2.7 (0.2)	0.9963	≤0.05	3
Ni	−50.03 (0.06)	−4.51 (0.01)	1	≤0.01	3

$$\log K_{sp}°(\text{ZnPo(s)}, (3.95)) = -35.5 \pm 0.5$$

$$\log K_{sp}°(\text{NiPo(s)}, (3.95)) = -42.4 \pm 0.5$$

where the uncertainties have been assigned in this review.

Use of these solubilities and the $\Delta G_f°$ values for Po^{2-}, Pb^{2+}, Hg^{2+}, Zn^{2+}, and Ni^{2+} gives the following $\Delta G_f°$ values (kJ mol^{-1}) for the metal polonides:

$$\Delta G_f°(\text{PbPo(s)}) = -(119.6 \pm 8.2)\ \text{kJ mol}^{-1}$$

$$\Delta G_f°(\text{HgPo(s)}) = -(57.1 \pm 8.2)\ \text{kJ mol}^{-1}$$

$$\Delta G_f°(\text{ZnPo(s)}) = -(152.8 \pm 8.2)\ \text{kJ mol}^{-1}$$

$$\Delta G_f°(\text{NiPo(s)}) = -(89.6 \pm 8.3)\ \text{kJ mol}^{-1}$$

The fit for Ag was not significant. However, Eq. (3.97) is not entirely appropriate to describe the solubility of the silver chalcogenides. For silver, the solubility data were modified to be related to Eq. (3.98):

$$\frac{1}{2}\text{Ag}_2\text{X(s)} + \text{H}^+ \rightleftharpoons \text{Ag}^+ + \frac{1}{2}\text{H}_2\text{X(aq)} \tag{3.98}$$

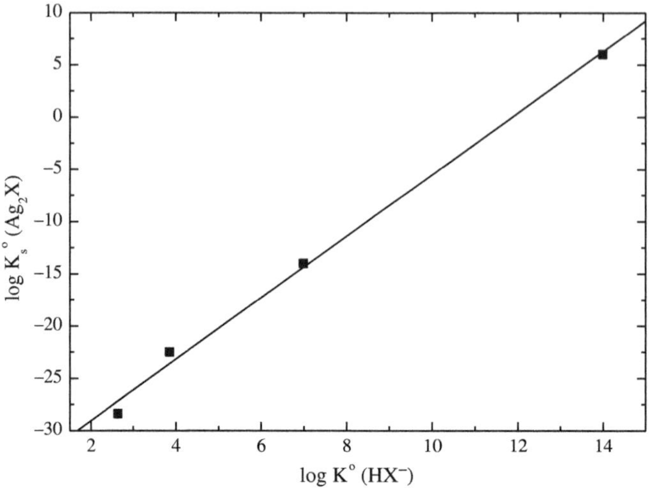

Figure 3.32 Plots of log $K°$ (HX^-) versus log $K_s°$ (Ag_2X) for silver chalcogenides.

where X is the chalcogen (O, S, Se, Te, or Po). The solubility data for Eq. (3.96) determined from the data listed in Table 3.9 and the relevant Gibbs energy data given in Appendix 1 are log $K_s = -28.4$, -22.5, and -14.0 for $Ag_2Te(s)$, $Ag_2Se(s)$, and $Ag_2S(s)$, respectively. In addition, the log K_s for $Ag_2O(s)$ is 5.99 (Brown and Ekberg, 2016).

A similar correlation as shown in Fig. 3.31 can now be carried out for silver (Fig. 3.32). The line of best fit has a slope of 2.94 ± 0.14 and an intercept of -34.9 ± 1.1 [$r^2 = 0.9956$; $P \leq 0.01$; $n = 4$] (the association constant for water is listed in Table 3.3). From this correlation, and using the association constant for HPo^- (log $K = 1.69$), the calculated solubility constant for Eq. (3.98) (X = Po) is:

$$\log K_s°(Ag_2Po(s), (3.96)) = -30.0 \pm 0.5$$

where the uncertainty has been assigned by the present review.

Use of this solubility and the $\Delta G_f°$ values for $H_2Po(aq)$, Ag^+, and H^+ gives the following $\Delta G_f°$ value (kJ mol^{-1}) for silver polonide:

$$\Delta G_f°(Ag_2Po(s)) = -(77.4 \pm 9.6) \text{ kJ mol}^{-1}$$

For comparison with the data listed in Table 3.9 for silver chalcogenide solubilities, the calculated solubility for silver polonide can be written with respect to Eq. (3.99):

$$Ag_2Po(s) \rightleftharpoons 2Ag^+ + Po^{2-} \quad (3.99)$$

The calculated solubility constant for this equation is:

$$\log K_{sp}°(Ag_2Po(s)) = -75.1$$

using the previous solubility constant and the two $H_2Po(aq)$ dissociation constants listed in Section 3.2.10. The latter solubility constant indicates that, as expected, the polonide of silver is less soluble than those solids with the lighter chalcogenides.

3.2.32 Other metal polonides

Krestov (1962) has calculated thermodynamic data for most of the alkali and alkaline earth polonides [data listed in Abakumov (1982)]. Gibbs energy, enthalpy, and entropy of formation data have been calculated for Na, K, Rb, Cs, Mg, Ca, Sr, Ba, and Ra polonides. However, for many of the metals, the calculated Gibbs energy and enthalpy data are the same value. This behavior is considered unrealistic and, as such, these data are not accepted by the present study.

3.2.33 $(PoO)_2(NO_3)_3OH(s)$

Orban (1947) studied the solubility of polonium(IV) nitrate. Bagnall et al. (1958) prepared a polonium(IV) nitrate by exposing either polonium hydroxide or tetrachloride to 0.5 mol L^{-1} HNO_3 for 12 hours. They concluded that the most likely structure of the nitrate was $(PoO)_2(NO_3)_3OH(s)$. From the measurements of Orban (1947) [data given in Moyer (1956)], a plot of the logarithm of the acid concentration against the logarithm of the solubility at high nitric acid concentrations (solid line) has a slope of 1.58 ± 0.09 [$r^2 = 0.9876$; $P \leq 0.001$; $n = 6$] (Fig. 3.33). This indicates that three moles of acid will solubilize two moles of polonium and is consistent with both the structure proposed by Bagnall et al. (1958) and the Po-NO_3 complexes indicated by Ampelogova (1973) (see Section 3.2.34). The equation can be written as follows:

$$(PoO)_2(NO_3)_3OH(s) + 3NO_3^- + H^+ \rightleftharpoons 2PoO(NO_3)_3^- + H_2O \quad (3.100)$$

The solubility has been described by this single reaction. However, at lower nitric acid concentrations the slope is lower and indicates a change in speciation in solution, most likely due to the formation of both PoO $(NO_3)_2(aq)$ and $PoO(NO_3)_3^-$, similar to the solubility behavior of Po

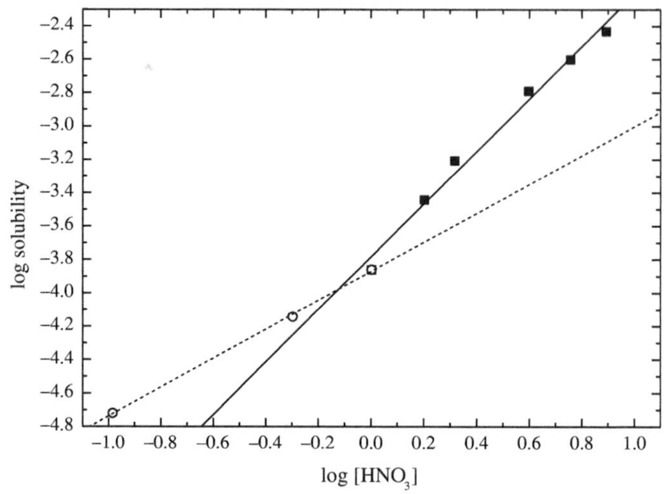

Figure 3.33 Plot of log [HNO_3] versus log solubility for $(PoO)_2(NO_3)_3OH(s)$.

$(SO_4)_2 \cdot H_2O(s)$. The slope of the dotted line in Fig. 3.33 is 0.87 ± 0.02 [$r^2 = 0.9993$; $P \leq 0.05$; $n = 3$].

A solubility constant of:

$$\log K°\left((PoO)_2(NO_3)_3OH(s)\right) = -7.17 \pm 0.05$$

for has been derived for Eq. (3.100) using the specific ion interaction theory (Grenthe et al., 1992) (Fig. 3.34). Use of this log $K°$ and the $\Delta G_f°$ values for $PoO(NO_3)_3^-$, NO_3^-, and H_2O gives:

$$\Delta G_f°\left((PoO)_2(NO_3)_3OH(s)\right) = -(502 \pm 11)\ \text{kJ mol}^{-1}$$

In turn, a solubility constant of:

$$\log K_{sp}°\left((PoO)_2(NO_3)_3OH(s)\right) = -25.8$$

has been derived for reaction (3.101) using this $\Delta G_f°$ value and those for PoO^{2+}, NO_3^-, and OH^-.

$$(PoO)_2(NO_3)_3OH(s) \rightleftharpoons 2PoO^{2+} + OH^- + 3NO_3^- \quad (3.101)$$

3.2.34 $PoONO_3^+$, $PoO(NO_3)_2(aq)$, and $PoO(NO_3)_3^-$

Ampelogova (1973) studied the complexation of polonium(IV) by nitrate using solvent extraction at ambient temperature. The stability constants (log β_n) determined for the species in Eqs. (3.102), (3.103), and (3.104)

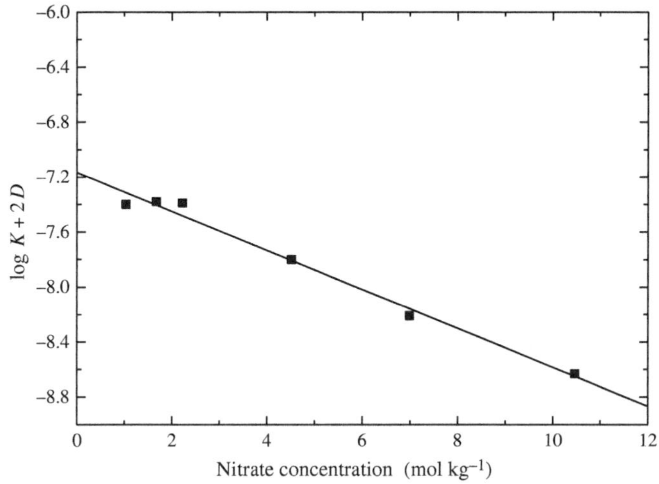

Figure 3.34 Plot of ionic strength versus log $K + 2D$ for $(PoO)_2(NO_3)_3OH(s)$.

were 0.56, 1.15, and 1.30 in 1 mol L^{-1} and 0.53, 1.08, and 1.30 in 1.5 mol L^{-1} $H(ClO_4, NO_3)$, respectively.

$$PoO^{2+} + NO_3^- \rightleftharpoons PoONO_3^+ \quad (3.102)$$

$$PoO^{2+} + 2NO_3^- \rightleftharpoons PoO(NO_3)_2(aq) \quad (3.103)$$

$$PoO^{2+} + 3NO_3^- \rightleftharpoons PoO(NO_3)_3^- \quad (3.104)$$

Correction of these data to zero ionic strength using the specific ion interaction theory (Grenthe et al., 1992) results in log $\beta_n°$ values of:

$$\log \beta_1°\left(PoONO_3^+\right) = 1.30 \pm 0.10$$

$$\log \beta_2°\left(PoO(NO_3)_2(aq)\right) = 2.32 \pm 0.10$$

$$\log \beta_3°\left(PoO(NO_3)_3^-\right) = 2.34 \pm 0.10$$

for Eqs. (3.102)−(3.104), respectively. As there were only two data for each species, the uncertainties have been assigned by the present review.

On the basis of these corrected log $\beta_n°$ values and the $\Delta G_f°$ values for PoO^{2+} and NO_3^-, the calculated $\Delta G_f°$ values of $PoONO_3^+$, PoO $(NO_3)_2(aq)$, and $PoO(NO_3)_3^-$ are:

$$\Delta G_f°(PoONO_3^+) = -(50.7 \pm 5.2)\ kJ\ mol^{-1}$$

$$\Delta G_f°(PoO(NO_3)_2(aq)) = -(167.3 \pm 5.3)\ kJ\ mol^{-1}$$

$$\Delta G_f°(PoO(NO_3)_3^-) = -(278.2 \pm 5.3)\ kJ\ mol^{-1}$$

3.2.35 $Po(CN)_6^{2-}$

By studying the relationship between the stability constants (log $\beta_n°$) for metal chloride and cyanide species, it may be possible to estimate a corresponding polonium cyanide value as described by Eq. (3.105).

$$PoO^{2+} + 6CN^- + 2H^+ \rightleftharpoons Po(CN)_6^{2-} + H_2O \quad (3.105)$$

The log $\beta_n°$ values for various metal ion chloride and cyanide species obtained from Sillén and Martell (1964, 1971) and Högfeldt (1982) are summarized in Table 3.11, with the corresponding relationship being represented graphically in Fig. 3.35.

The line of best fit in Fig. 3.35 has an intercept of 4.5 (0.6) and a slope of 1.80 ± 0.16 [$r^2 = 0.9635$; $P \le 0.001$; $n = 7$]. From the previously determined value of 14.55 for the log $\beta_6°$ of $PoCl_6^{2-}$ and the line of best fit from Fig. 3.35, a stability constant of:

Table 3.11 log $\beta_n°$ values of metal chloride and cyanide complexes at 25°C and zero ionic strength.

Metal ion	n	log $\beta_n°$ [a]	
		$MCl_n^{(n-z)-}$	$M(CN)_n^{(n-z)-}$
Ag(I)	2	5.18	19.81
Cu(I)	2	5.50	21.30
Hg(II)	2	14.00	34.71
Cd(II)	4	1.6	17.92
Pd(II)	4	15.70	42.4
Pt(II)	4	14.4	41.4
Zn(II)	4	0.20	19.62

[a]z = the ionic charge of the metal.
Source: From Sillén, L.G. and Martell, A.E., 1964. Stability Constants of Metal-Ion Complexes. The Chemical Society, London, Special Publication No. 17, Sillén, L.G. and Martell, A.E., 1971. Stability Constants of Metal–Ion Complexes. The Chemical Society, London, Special Publication No. 25, and Högfeldt, E., 1982. Stability Constants of Metal–Ion Complexes. Part A: Inorganic Ligands. Pergamon Press, Oxford.

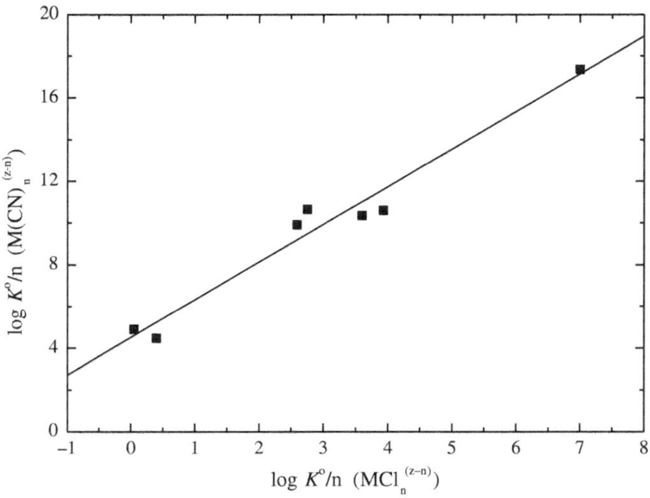

Figure 3.35 Relationship between the log $\beta_n°$ for metal chloride and cyanide complexes.

$$\log \beta_6°\left(\mathrm{Po(CN)_6^{2-}}\right) = 53.3 \pm 0.3$$

for the formation of $Po(CN)_6^{2-}$ in Eq. (3.105) has been calculated. The uncertainty has been assigned in this review. The $\Delta G_f°$ value for $Po(CN)_6^{2-}$ subsequently derived using this log $\beta_6°$ and the $\Delta G_f°$ values for PoO^{2+}, CN^- and H_2O is:

$$\Delta G_f°\left(\mathrm{Po(CN)_6^{2-}}\right) = (1002 \pm 16)\ \mathrm{kJ\ mol^{-1}}$$

The uncertainty has been assigned by the present review.

3.2.36 $PoO(CN)_2$(s)

Bagnall (1957) studied the solubility of polonium cyanide in 0.02–1.5 mol L^{-1} potassium cyanide at 22°C while Moyer (1956) gives a value of 3.8 mCi mL^{-1} for the solubility of oxidized polonium (presumably polonium dioxide) in 1.0 mol L^{-1} KCN at ambient temperature. From their measurements, a plot of the logarithm of the acid concentration versus the logarithm of the solubility (Fig. 3.36) shows that there is a change in solution speciation at the point where the two lines on the figure cross.

The line of best fit for the portion of the graph shown by the solid line has a slope of 1.87 ± 0.08 [$r^2 = 0.9980$; $P \leq 0.05$; $n = 3$], and

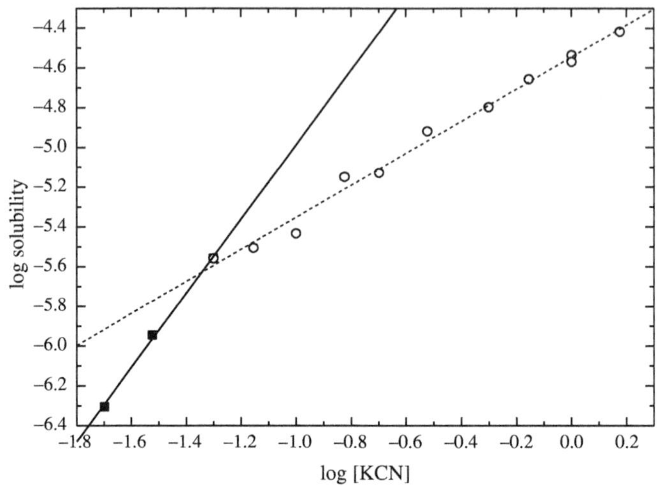

Figure 3.36 Plot of log [KCN] versus log solubility for $PoO(CN)_2(s)$.

therefore one mole of polonium will be solubilized by two moles of cyanide. The reaction can be described by Eq. (3.106).

$$2PoO(CN)_2(s) + 2CN^- \rightleftharpoons PoO_2(s) + Po(CN)_6^{2-} \quad (3.106)$$

The intercept of the line is equivalent to the solubility constant for Eq. (3.106).

$$\log K^\circ\left(PoO(CN)_2(s)\right) = -3.12 \pm 0.13$$

Use of this log K° and the ΔG_f° values for $Po(CN)_6^{2-}$, $PoO_2(s)$ and CN^- gives:

$$\Delta G_f^\circ\left(PoO(CN)_2(s)\right) = (229.1 \pm 8.6)\ \text{kJ mol}^{-1}$$

for the Gibbs energy of $PoO(CN)_2(s)$.

The line of best fit for the portion of the graph shown by the dotted line has a slope of 0.81 ± 0.03 [$r^2 = 0.9904$; $P \leq 0.001$; $n = 11$], and therefore two moles of polonium will be solubilized by two moles of cyanide. The reaction can be described by Eq. (3.107).

$$2PoO(CN)_2(s) + 2CN^- + OH^- \rightleftharpoons Po(CN)_6^{2-} + HPoO_3^- \quad (3.107)$$

A ΔG_r of 26.0 kJ mol^{-1} has been determined for Eq. (3.107) using the ΔG_f° values for $HPoO_3^-$, $Po(CN)_6^{2-}$, $PoO(CN)_2(s)$, CN^-, and OH^-. A solubility constant of:

$$\log K^{\circ}\left(\mathrm{PoO(CN)_2(s)}, (3.105)\right) = -4.55 \pm 0.02$$

has been calculated for Eq. (3.107) from this ΔG_r value.

For Eq. (3.108),

$$\mathrm{PoO(CN)_2(s) \rightleftharpoons PoO^{2+} + 2CN^-} \quad (3.108)$$

a solubility constant of

$$\log K^{\circ}\left(\mathrm{PoO(CN)_2(s)}\right) = -29.3$$

has been calculated.

3.2.37 Organic complexes of polonium

Ampelogova (1973) studied the complexation of polonium(IV) by acetate using solvent extraction at ambient temperature. The stability constants (log β_n) determined for the species in Eqs. (3.109)–(3.112) were 2.51, 4.85, 7.18, and 8.0 in 1 mol L^{-1} $NaClO_4$. On the basis of the trend in the stability constants (i.e., the stepwise constant K_3 is virtually equal to K_2 when it would be expected to be somewhat smaller), there would appear to be a change in the stoichiometry of the reactions. The following reactions are proposed on the basis of the data:

$$\mathrm{PoO^{2+} + CH_3COO^- \rightleftharpoons PoOCH_3COO^+} \quad (3.109)$$

$$\mathrm{PoO^{2+} + 2CH_3COO^- \rightleftharpoons PoO(CH_3COO)_2(aq)} \quad (3.110)$$

$$\mathrm{PoO^{2+} + 3CH_3COO^- + 2H^+ \rightleftharpoons Po(CH_3COO)_3^+ + H_2O} \quad (3.111)$$

$$\mathrm{PoO^{2+} + 4CH_3COO^- + 2H^+ \rightleftharpoons Po(CH_3COO)_4(aq) + H_2O} \quad (3.112)$$

Correction of these data to zero ionic strength using the Davies equation (Davies, 1962) results in log $\beta_n{}^{\circ}$ values for Eqs. (3.109)–(3.112) of:

$$\log \beta_1{}^{\circ}\left(\mathrm{PoOCH_3COO^+}\right) = 2.69 \pm 0.30$$

$$\log \beta_2{}^{\circ}\left(\mathrm{PoO(CH_3COO)_2(aq)}\right) = 5.12 \pm 0.30$$

$$\log \beta_3{}^{\circ}\left(\mathrm{Po(CH_3COO)_3^+}\right) = 7.54 \pm 0.30$$

$$\log \beta_4°(Po(CH_3COO)_4(aq)) = 8.45 \pm 0.30$$

The uncertainties have been assigned by the present review.

On the basis of these corrected log β_n° values and the ΔG_f° values for PoO^{2+}, H^+, and CH_3COO^-, the calculated ΔG_f° values of $PoOCH_3COO^+$, $PoO(CH_3COO)_2(aq)$, $Po(CH_3COO)_3^+$, and Po $(CH_3COO)_4(aq)$ are:

$$\Delta G_f°(PoOCH_3COO^+) = -(317.2 \pm 5.5)\ \text{kJ mol}^{-1}$$

$$\Delta G_f°(PoO(CH_3COO)_2(aq)) = -(700.5 \pm 5.5)\ \text{kJ mol}^{-1}$$

$$\Delta G_f°(Po(CH_3COO)_3^+) = -(846.5 \pm 5.6)\ \text{kJ mol}^{-1}$$

$$\Delta G_f°(Po(CH_3COO)_4(aq)) = -(1221.1 \pm 5.8)\ \text{kJ mol}^{-1}$$

As will be shown below with the complexation behavior of polonium (IV) with other organic ligands, the oxygen in the PoO molecule seems to be retained when bound by two or less organic ligands, but is lost when three or more ligands are complexed.

Bagnall (1957) studied the solubility of polonium acetate in 0.1–2.0 mol L^{-1} acetic acid at 22°C while Moyer (1956) gives a value of 128 mCi mL^{-1} for the solubility of oxidized polonium (presumably polonium dioxide) in 1.0 mol L^{-1} acetic acid at ambient temperature. From their measurements, a plot of the logarithm of the acid concentration versus the logarithm of the solubility (Fig. 3.37) is linear across the full range of concentrations of acid studied.

The line of best fit has a slope of 2.20 ± 0.07 [$r^2 = 0.9874$; $P \leq 0.001$; $n = 13$], and therefore one mole of polonium is solubilized by two moles of acetate. The reaction can be described by Eq. (3.113).

$$PoO(CH_3COO)_2(s) + 2CH_3COOH \rightleftharpoons Po(CH_3COO)_4(aq) + H_2O \tag{3.113}$$

The intercept of the line is equivalent to the solubility constant for Eq. (3.113).

$$\log K°(PoO(CH_3COO)_2(s)) = -3.14 \pm 0.03$$

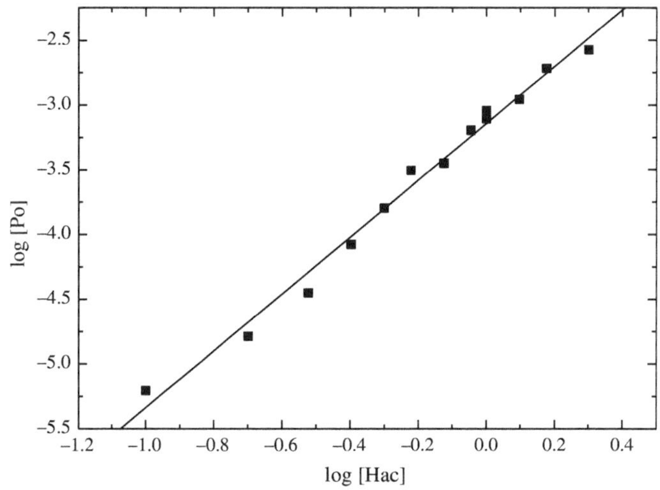

Figure 3.37 Plot of log $[CH_3COOH]$ versus log solubility for $PoO(CH_3COO)_2(s)$.

Use of this log K° and the ΔG_f° values for $Po(CH_3COO)_4(aq)$, $CH_3COOH(aq)$ and H_2O gives:

$$\Delta G_f^\circ\left(PoO(CH_3COO)_2(s)\right) = -(677.0 \pm 5.9)\ \text{kJ mol}^{-1}$$

for the Gibbs energy of $PoO(CH_3COO)_2(s)$.

Koch and Falkenburg (1967) studied the complexation of polonium (IV), using solvent extraction, with a number of organic ligands employing thenoyltrifluoroacetone (TTA) as the organic phase extractant. They were able to determine the stability constants of the predominant polonium(IV)–organic ligand complex as well as that of the aqueous polonium(IV)–TTA complexes. Eq. (3.114) describes the complexation reactions:

$$PoO^{2+} + qL^{m-} \rightleftharpoons PoOL_q^{(2-qm)} \tag{3.114}$$

In general, the charge on the complex formed was -2. The organic ligands studied were TTA, oxalate, tartrate, citrate, nitrilotriacetic acid, and ethylenediamine tetraacetic acid. The study was conducted at 22°C and using 1 mol L^{-1} sodium perchlorate as the medium. The stability constants obtained for Eq. (3.114) were:

$$\log K\left(PoOTTA^+, q = 1, m = 1\right) = 6.6$$

Table 3.12 Selected thermochemical data for polonium species at 101 kPa and 298.15K.

Species	Reaction	Parameter	Value	Comment/Reference
Po(s)		ΔG_f°	0.0	Standard state
		ΔH_f°	0.0	Standard state
		S_f°	62.8 ± 2.1	Stull and Sinke (1956)
Po(g)		ΔG_f°	106.6 ± 0.2	
		ΔH_f°	144.1 ± 0.2	Stull and Sinke (1956)
		S_f°	188.80 ± 0.04	
$Po_2(g)$		ΔG_f°	93.0 ± 0.4	
		ΔH_f°	137.7 ± 0.4	Stull and Sinke (1956)
		S_f°	275.3 ± 0.1	
Po^{2-}	$Po(s) + 2e^- \rightleftharpoons Po^{2-}$	ΔG_f°	197.1 ± 7.7	Calculated from E°
		E°	− 1.02	Electrostatic correlation
Po^{2+}	$Po^{2+} + 2e^- \rightleftharpoons Po(s)$	ΔG_f°	124.1 ± 8.4	Calculated from E°
		E°	0.643 ± 0.037	Average of literature data
$PoO_2(s)$	$PoO_2(s) + 4H^+ + 4e^- \rightleftharpoons Po(s) + 2H_2O$	ΔG_f°	− 192.1 ± 3.3	Calculated from E°
		E°	0.731 ± 0.009	Average of literature
		ΔH_f°	− 251	Zhdanov (1985)
		S_f°	71	Zhdanov (1985)
$H_2PoO_3(aq)$	$PoO_2(s) + H_2O \rightleftharpoons H_2PoO_3(aq)$	ΔG_f°	− 392.5 ± 3.8	
	$HPoO_3^- + H^+ \rightleftharpoons H_2PoO_3(aq)$	log K°	12.41 ± 0.30	Calculated from ΔG_f° values
${PoO_3}^{2-}$	$PoO_2(s) + 2OH^- \rightleftharpoons PoO_3^{2-} + H_2O$	ΔG_f°	− 244.2 ± 3.4	Calculated from log K°
		log K°	− 4.43 ± 0.02	
${HPoO_3}^-$		ΔG_f°	− 321.6 ± 4.1	Calculated from log K°
	$PoO_2(s) + OH^- \rightleftharpoons HPoO_3^-$	log K°	− 5.01 ± 0.51	
	$PoO_3^{2-} + H^+ \rightleftharpoons HPoO_3^-$	log K°	13.56 ± 0.30	Calculated from ΔG_f° values
$PoOOH^+$	$H_2PoO_3(aq) + H^+ \rightleftharpoons PoOOH^+ + H_2O$	ΔG_f°	− 164.8 ± 4.9	Calculated from log K°
		log K°	1.7 ± 0.6	

PoO^{2+}	$PoO^{2+} + H_2O \rightleftharpoons HPoO_3^- + H^+$	ΔG_f°	67.6 ± 5.2	Calculated from log K°
		log K°	-0.85 ± 0.30	
		ΔG_f°	-138.1 ± 3.1	Calculated from E° values
$PoO_3(s)$	$PoO_3(s) + 2H^+ + 2e^- \rightleftharpoons PoO_2(s) + H_2O$	E°	1.509	Zhdanov (1985)
		E°	0.55	Haissinsky (1946)
	$PoO_3(s) + 2e^- \rightleftharpoons PoO_3^{2-}$			
HPo^-	$HPo^- \rightleftharpoons Po^{2-} + H^+$	ΔG_f°	120.4 ± 7.0	Electrostatic correlation
		log K°	-13.44	Calculated from ΔG_f° values
$H_2Po(aq)$	$H_2Po(aq) \rightleftharpoons HPo^- + H^+$	ΔG_f°	110.8 ± 7.7	Electrostatic correlation
		log K°	-1.69	Calculated from ΔG_f° values
$PoCl_2(s)$		ΔG_f°	-136 ± 14	Linear free energy
		S_f°	130	Zhdanov (1985)
		ΔG_f°	-243.0 ± 4.7	Linear free energy relationship
$PoCl_4(s)$		S_f°	197	Zhdanov (1985)
		ΔG_f°	-444.0 ± 2.3	Calculated from E°
	$PoCl_4^{2-} + 2e^- \rightleftharpoons Po(s) + 4Cl^-$	E°	0.419 ± 0.012	SIT analysis of literature data
$PoCl_4{}^{2-}$	$Po^{2+} + 4Cl^- \rightleftharpoons PoCl_4^{2-}$	log K°	7.58	Calculated from ΔG_f° values
$PoCl_3{}^-$	$Po^{2+} + 3Cl^- \rightleftharpoons PoCl_3^-$	ΔG_f°	-308.5 ± 8.4	Calculated from log K°
		log K°	6.84 ± 0.20	Linear free energy relationship
$PoCl_2(aq)$	$Po^{2+} + 2Cl^- \rightleftharpoons PoCl_2(aq)$	ΔG_f°	-172.2 ± 8.4	Calculated from log K°
		log K°	5.95 ± 0.20	Linear free energy relationship
$PoCl^+$	$Po^{2+} + Cl^- \rightleftharpoons PoCl^+$	ΔG_f°	-27.4 ± 8.4	Calculated from log K°
		log K°	3.56 ± 0.20	Linear free energy relationship
		ΔG_f°	-565.6 ± 0.5	Calculated from E°
	$PoCl_6^{2-} + 2e^- \rightleftharpoons PoCl_4^{2-} + 2Cl^-$			
	$PoO^{2+} + 6Cl^- + 2H^+$	E°	0.730 ± 0.001	SIT analysis of literature data
$PoCl_6{}^{2-}$	$\rightleftharpoons PoCl_6^{2-} + H_2O$	log K°	14.55	Calculated from ΔG_f° values
	$PoCl_6^{2-} + 4e^- \rightleftharpoons Po(s) + 6Cl^-$	E°	0.574	Calculated from ΔG_f° values

(Continued)

Table 3.12 (Continued)

Species	Reaction	Parameter	Value	Comment/Reference
$PoOHCl_4^-$	$PoOHCl_4^- + 2Cl^- + H^+ \rightleftharpoons PoCl_6^{2-} + H_2O$	ΔG_f^o	-527.5 ± 0.7	Calculated from log K^o
		log K^o	2.25 ± 0.07	
$PoOCl_4^{2-}$	$PoOCl_4^{2-} + 2H_2O \rightleftharpoons H_2PoO_3(aq) + 4Cl^- + 2H^+$	ΔG_f^o	-493.8 ± 6.3	Calculated from log K^o
		log K^o	-8.9	
	$PoO^{2+} + 4Cl^- \rightleftharpoons PoOCl_4^{2-}$	log K^o	6.4 ± 0.3	Calculated from ΔG_f^o values
$PoOOHCl_2^-$	$PoOOHCl_2^- + H_2O \rightleftharpoons H_2PoO_3(aq) + 2Cl^- + H^+$	ΔG_f^o	-445.1 ± 6.3	Calculated from log K^o
		log K^o	-4.8	
$PoBr_2(s)$		ΔG_f^o	-52.2 ± 5.9	Linear free energy relationship
		S_f^o	155	Zhdanov (1985)
$PoBr_4(s)$		ΔG_f^o	-153.7 ± 5.3	Linear free energy relationship
		S_f^o	230	Zhdanov (1985)
$PoI_2(s)$		ΔG_f^o	95.2 ± 4.5	Linear free energy relationship
$PoI_4(s)$		ΔG_f^o	95.2 ± 4.5	Linear free energy relationship
PoI_5^-	$PoI_4(s) + I^- \rightleftharpoons PoI_5^-$	ΔG_f^o	-74.6 ± 7.3	Calculated from log K^o
		log K^o	-4.17	
PoI_6^{2-}	$PoI_4(s) + 2I^- \rightleftharpoons PoI_6^{2-}$	ΔG_f^o	-137.4 ± 7.3	Calculated from log K^o
		log K^o	-2.23	
$PoS(s)$	$PoS(s) \rightleftharpoons Po^{2+} + S^{2-}$	ΔG_f^o	61.8 ± 8.7	Calculated from log K^o
		log K^o	-27.26	
$PoSO_4(s)$	$PoSO_4(s) \rightleftharpoons Po^{2+} + SO_4^{2-}$	ΔG_f^o	-670.7 ± 8.6	Calculated from log K^o
		log K^o	-8.90	
$PoOSO_4(aq)$	$PoO^{2+} + SO_4^{2-} \rightleftharpoons PoOSO_4(aq)$	ΔG_f^o	-691.1 ± 5.4	Calculated from log K^o
		log K^o	2.56 ± 0.30	
$PoO(SO_4)_2^{2-}$	$PoO^{2+} + 2SO_4^{2-} \rightleftharpoons PoO(SO_4)_2^{2-}$	ΔG_f^o	-1441.1 ± 5.5	Calculated from log K^o

$PoO(SO_4)_3{}^{4-}$	$PoO(SO_4)_2^{2-} + SO_4^{2-} \rightleftharpoons PoO(SO_4)_3^{4-}$	$\log K°$	3.61 ± 0.30	
		$\Delta G_f°$	-2189.3 ± 5.7	Calculated from $\log K°$
$Po(SO_4)_2 \cdot H_2O(s)$	$Po(SO_4)_2 \cdot H_2O(s) + HSO_4^- \rightleftharpoons PoO(SO_4)_3^{4-} + 3H^+$	$\log K°$	0.75 ± 0.24	
		$\Delta G_f°$	-1487.9 ± 5.8	Calculated from $\log K°$
		$\log K°$	-9.44 ± 0.06	SIT analysis of literature data
$(PoO)_2OSO_4(s)$		$\Delta G_f°$	-907.6 ± 8.0	Calculated from $\log K°$
	$(PoO)_2OSO_4(s) + HSO_4^- + H^+ \rightleftharpoons 2PoOSO_4(aq) + H_2O$	$\log K°$	-7.75 ± 0.08	SIT analysis of literature data
	$(PoO)_2OSO_4(s) + H_2O \rightleftharpoons 2PoO^{2+} + 2OH^- + SO_4^{2-}$	$\log K°$	-38.8 ± 0.1	Calculated from $\Delta G_f°$ values
$PoSO_4(aq)$		$\Delta G_f°$	-626.3 ± 3.9	Calculated from $E°$
	$PoSO_4(aq) + 2e^- \rightleftharpoons Po(s) + SO_4^{2-}$	$E°$	0.61 ± 0.02	Calculated from $\Delta G_f°$ values
	$Po^{2+} + SO_4^{2-} \rightleftharpoons PoSO_4(aq)$	$\log K°$	1.13	
$(PoO)_2OSeO_4(s)$		$\Delta G_f°$	-605.7 ± 8.6	Calculated from $\log K°$ values
	$(PoO)_2OSeO_4(s) + HSeO_4^- + H^+ \rightleftharpoons 2PoOSeO_4(aq) + H_2O$	$\log K°$	-8.53 ± 0.09	SIT analysis of literature data
	$(PoO)_2OSeO_4(s) + 3HSeO_4^- \rightleftharpoons 2PoO(SeO_4)_2^{2-} + H_2O + H^+$	$\log K°$	-10.82 ± 0.11	SIT analysis of literature data
	$(PoO)_2OSeO_4(s) + H_2O \rightleftharpoons 2PoO^{2+} + 2OH^- + SeO_4^{2-}$	$\log K°$	-39.2 ± 0.4	Calculated from $\Delta G_f°$ values
$PoOSeO_4(aq)$	$PoO^{2+} + SeO_4^{2-} \rightleftharpoons PoOSeO_4(aq)$	$\Delta G_f°$	-385.8 ± 5.6	Calculated from $\log K°$
		$\log K°$	2.43 ± 0.30	
$PoO(SeO_4)_2{}^{2-}$		$\Delta G_f°$	-826.1 ± 6.4	Calculated from $\log K°$
		$\log K°$	2.57 ± 0.42	
$Ag_2PoO_3(s)$	$Ag_2PoO_3(s) \rightleftharpoons 2Ag^+ + PoO_3^{2-}$	$\Delta G_f°$	-231.7 ± 3.6	Calculated from $\log K°$
		$\log K°$	-24.8	Linear free energy relationship
$PbPo(s)$	$PbPo(s) \rightleftharpoons Pb^{2+} + Po^{2-}$	$\Delta G_f°$	-119.6 ± 8.2	Calculated from $\log K°$
		$\log K°$	-51.2 ± 0.5	Linear free energy relationship

(Continued)

Table 3.12 (Continued)

Species	Reaction	Parameter	Value	Comment/Reference
$HgPo(s)$	$HgPo(s) \rightleftharpoons Hg^{2+} + Po^{2-}$	ΔG_f°	-57.1 ± 8.2	Calculated from log K°
		log K°	-73.4 ± 0.5	Linear free energy relationship
$ZnPo(s)$	$ZnPo(s) \rightleftharpoons Zn^{2+} + Po^{2-}$	ΔG_f°	-152.8 ± 8.2	Calculated from log K°
		log K°	-35.5 ± 0.5	Linear free energy relationship
$NiPo(s)$	$NiPo(s) \rightleftharpoons Ni^{2+} + Po^{2-}$	ΔG_f°	-89.6 ± 8.3	Calculated from log K°
		log K°	-42.4 ± 0.5	Linear free energy relationship
		ΔG_f°	-77.4 ± 9.6	Calculated from log K°
$Ag_2Po(s)$	$\frac{1}{2}Ag_2Po(s) + H^+ \rightleftharpoons Ag^+ + \frac{1}{2}H_2Po(aq)$	log K°	-30.0 ± 0.5	Linear free energy relationship
	$Ag_2Po(s) \rightleftharpoons 2Ag^+ + Po^{2-}$	log K°	-75.1	Calculated from ΔG_f° values
		ΔG_f°	-502 ± 11	Calculated from log K°
$(PoO)_2(NO_3)_3OH(s)$	$(PoO)_2(NO_3)_3OH(s) + 3NO_3^- + H^+ \rightleftharpoons 2PoO(NO_3)_3^- + H_2O$	log K°	-7.17 ± 0.05	SIT analysis of literature data
	$(PoO)_2(NO_3)_3OH(s) \rightleftharpoons 2PoO^{2+} + OH^- + 3NO_3^-$	log K°	-25.8	Calculated from ΔG_f° values
$PoONO_3^+$	$PoO^{2+} + NO_3^- \rightleftharpoons PoONO_3^+$	ΔG_f°	-50.7 ± 5.2	Calculated from log K°
		log K°	1.30 ± 0.10	
$PoO(NO_3)_2(aq)$	$PoO^{2+} + 2NO_3^- \rightleftharpoons PoO(NO_3)_2(aq)$	ΔG_f°	-167.3 ± 5.3	Calculated from log K°
		log K°	2.32 ± 0.10	
$PoO(NO_3)_3^-$	$PoO^{2+} + 3NO_3^- \rightleftharpoons PoO(NO_3)_3^-$	ΔG_f°	-278.2 ± 5.3	Calculated from log K°
		log K°	2.34 ± 0.10	
$Po(CN)_6^{2-}$	$PoO^{2+} + 6CN^- + 2H^+ \rightleftharpoons Po(CN)_6^{2-} + H_2O$	ΔG_f°	1002 ± 16	Calculated from log K°
		log K°	53.3 ± 0.3	Linear free energy relationship
		ΔG_f°	229.1 ± 8.6	Calculated from log K°

$PoO(CN)_2(s)$	$2PoO(CN)_2(s) + 2CN^- \rightleftharpoons PoO_2(s) + Po(CN)_6^{2-}$	$\log K°$	-3.12 ± 0.13	
	$2PoO(CN)_2(s) + 2CN^- + OH^- \rightleftharpoons HPoO_3^- + Po(CN)_6^{2-}$	$\log K°$	-4.55 ± 0.02	
	$PoO(CN)_2(s) \rightleftharpoons PoO^{2+} + 2CN^-$	$\log K°$	-29.3	Calculated from $\Delta G_f°$ values
$PoOCH_3COO^+$	$PoO^{2+} + CH_3COO^- \rightleftharpoons PoOCH_3COO^+$	$\Delta G_f°$	-317.2 ± 5.5	Calculated from $\log K°$
		$\log K°$	2.69 ± 0.30	
$PoO(CH_3COO)_2(aq)$	$PoO^{2+} + 2CH_3COO^- \rightleftharpoons PoO(CH_3COO)_2(aq)$	$\Delta G_f°$	-700.5 ± 5.5	Calculated from $\log K°$
		$\log K°$	5.12 ± 0.30	
$Po(CH_3COO)_3{}^+$	$PoO^{2+} + 3CH_3COO^- + 2H^+ \rightleftharpoons Po(CH_3COO)_3^+ + H_2O$	$\Delta G_f°$	-846.5 ± 5.6	Calculated from $\log K°$
		$\log K°$	7.54 ± 0.30	
$Po(CH_3COO)_4(aq)$	$PoO^{2+} + 4CH_3COO^- + 2H^+ \rightleftharpoons Po(CH_3COO)_4(aq) + H_2O$	$\Delta G_f°$	-1221.1 ± 5.8	Calculated from $\log K°$
		$\log K°$	8.45 ± 0.30	
$PoO(CH_3COO)_2(s)$	$PoO(CH_3COO)_2(s) + 2CH_3COOH \rightleftharpoons Po(CH_3COO)_4(aq) + H_2O$	$\Delta G_f°$	-677.0 ± 5.9	Calculated from $\log K°$
		$\log K°$	-3.14 ± 0.03	

Notes: $\Delta G_f°$ is in kJ mol^{-1}; $\Delta H_f°$ is in kJ mol^{-1}; $S_f°$ is in J mol^{-1} K^{-1}; $E°$ in V.
SIT is the specific ion interaction theory.

$$\log K\big(\mathrm{PoO(TTA)_2(aq)}, q=2, m=1\big)=13.1$$

$$\log K\big(\mathrm{PoO(C_2O_4)_2^{2-}}, q=2, m=2\big)=7.8$$

$$\log K\big(\mathrm{PoOC_4H_4O_6^{2-}}, q=1, m=4\big)=7.3$$

$$\log K\big(\mathrm{PoO(Hcit)_2^{2-}}, q=2, m=2\big)=9.0$$

$$\log K\big(\mathrm{PoO(HNTA)_2^{2-}}, q=2, m=2\big)=8.2$$

$$\log K(\mathrm{PoOHEDTA^{-}}, q=1, m=3)=8.0$$

3.2.38 Summary of thermochemical data

Table 3.12 contains a summary of the calculated thermochemical data for polonium species.

References

Abakumov, A.S., 1982. Thermal reactions of polonium. Russ. Chem. Rev. 51, 622–629.

Abakumov, A.S., Ershova, Z.V., 1974. Vapor tension and thermal dissociation of polonium dioxide. Sov. Radiochem. 16, 401–404.

Ampelogova, N.I., 1973. Ion-exchange study of the complexing of polonium. Radiokhimiya 15, 813–820.

Aylward, G.H., Findlay, T.J.V., 1974. SI Chemical Data., second ed. John Wiley and Sons, Milton.

Baes, C.F., Mesmer, R.E., 1976. The Hydrolysis of Cations. John Wiley and Sons, New York.

Bagnall, K.W., 1957. Chemistry of the Rare Radioelements. Butterworths Scientific Publications, London.

Bagnall, K.W., 1983. The chemistry of polonium. Radiochim. Acta 32, 153–161.

Bagnall, K.W., Freeman, J.H., 1956a. Electrochemical studies on polonium. J. Chem. Soc. 2770–2774.

Bagnall, K.W., Freeman, J.H., 1956b. The sulphates and selenate of polonium. J. Chem. Soc. 4579–4582.

Bagnall, K.W., Freeman, J.H., 1957. Solubility of some polonium compounds. J. Chem. Soc. 2161–2163.

Bagnall, K.W., Robertson, D.S., 1957. Polonium monosulphide. J. Chem. Soc. 1044–1046.

Bagnall, K.W., D'Eye, R.W.M., Freeman, J.H., 1955. The polonium halides. Part I. Polonium chlorides. J. Chem. Soc. 2320–2326.

Bagnall, K.W., D'Eye, R.W.M., Freeman, J.H., 1956. The polonium halides. Part III. Polonium tetraiodide. J. Chem. Soc. 3385–3389.

Bagnall, K.W., Robertson, D.S., Stewart, M.A.A., 1958. The polonium nitrates. J. Chem. Soc. 3633–3636.
Bard, A.J., Parsons, R., Jordan, J. (Eds.), 1985. Standard Potentials in Aqueous Solution. Marcel Dekker Inc, New York.
Bent, H.E., French, C.L., 1941. The structure of ferric thiocyanate and its dissociation in aqueous solution. J. Am. Chem. Soc. 63, 568–572.
Bigelis, V.M., Makhkamova, M.Kh, Abrarov, O.A., 1978. Steady-state and equilibrium potentials of tellurium in iodide electrolytes. Electrokhimiya 14, 748–751.
Borzhim, V.S., Levchenko, V.I., Mogil'nitskii, A.A., Prokopovich, L.I., 1979. Determination of standard free energy of tellurium(IV) iodide. Ukr. Khim. Zh. 45, 789–790.
Brooks, L.S., 1955. The vapour pressure of polonium. J. Am. Chem. Soc. 77, 3211.
Brown, S.A., 2001. The Aqueous Chemistry of Polonium and its Relationship to Mineral Processing Streams (PhD dissertation). University of Western Sydney.
Brown, P.L., Ekberg, C., 2016. Hydrolysis of Metal Ions. Wiley-VCH, Weinheim, 917 p. (2 volumes).
Brown, P.L., Sylva, R.N., 1987. Unified Theory of Metal Ion Complex Formation Constants. Australian Atomic Energy Commission, AAEC/E656.
Brown, P.L., Ekberg, C., Ramebäck, H., Hedström, H., Matyskin, A., 2015. In: Merkel, B.J., Arab, A. (Eds.), Solubility of radium and strontium sulfate across the temperature range of 0 to 300 °C in Uranium – Past and Future Challenges. Springer, Heidelberg, pp. 553–563.
Buketov, E.A., Ugorets, M.Z., Pashinskin, A.S., 1964. Solubility product and entropy of sulphides, selenides and tellurides. Zh. Neorg. Khim 9, 526–529.
Charlot, G., 1958. Oxidation-Reduction Potentials. Pergammon Press, London.
Davies, C.W., 1962. Ion Association. Buttersworth Inc, Washington, DC.
Driessens, F.C.M., 1984. Solubility behaviour of ionic solids and their precipitation from aqueous solution. Bull. Soc. Chim. Belg. 93, 85–97.
Eberhart, J.G., McDonald, J.E., 1965. Graphical estimation of thermodynamic properties. J. Chem. Educ. 42, 601–603.
Eichelberger, J.F., Grove, G.R., Jones, L.V., 1965. Mound Laboratory Progress Report for March 1965. Mound Laboratory Report, MLM-1250.
Grenthe, I., Fuger, J., Konings, R.J.M., Lemire, R.J., Muller, A.B., Nguyen-Trung, C., et al., 1992. In: Wanner, H., Forest, I. (Eds.), Chemical Thermodynamics of Uranium. Elsevier, Amsterdam.
Haag, J., Alpen, J.V., Gmelin, E., Rabenau, A., 1979. Electrochemical and specific heat measurements on tellurium-halogen systems. Z. Naturforsch 34a, 969–975.
Haissinsky, M., 1932. Recherche électrochimique sur le polonium. J. Chim. Phys. 29, 453–473.
Haissinsky, M., 1946. Electrochimie des Substances Radioactives. Masson et Cie, Paris.
Haring, M.M., 1945. The Solubility of Polonium Oxide in Sodium Hydroxide. Mound Laboratory Report, MLM-45-9-43.
Harlow, B.B., 1947. Mound Laboratory Notes. MLM-M-76.
Hataye, I., Suganuma, H., Sakata, M., Nagame, Y., 1981. Solvent extraction study on the hydrolysis of tracer concentration of polonium(IV) in perchlorate solutions. J. Inorg. Nucl. Chem 43, 2101–2104.
Hisham, M.W.M., Benson, S.W., 1992. Thermochemistry of inorganic solids. 10. Empirical relations between the enthalpies of formation of solid halides and the corresponding gas-phase halide anions. J. Chem. Eng. Data 37, 194–199.
Högfeldt, E., 1982. Stability Constants of Metal-Ion Complexes. Part A: Inorganic Ligands. Pergamon Press, Oxford.

Hunt, D.J., 1954. Polonium Complexes in Chloride Solutions by Absorbancy Studies. Mound Laboratory Report, MLM-979.

Joliot, F., 1930. Étude électrochimque des radioéléments applications diverses. J. Chim. Phys. 27, 9−159.

Kapustinskii, A.F., 1948. Thermochemistry and the structure of atoms. Izvest. Akad. Nauk SSSR, Otdel. Khim. Nauk, pp. 568−580.

Katzlberger, C., 2000. Speciation, Analysis and Ion-Exchange Behaviour of Polonium and Other Natural Radionuclides in Drinking Water. PhD dissertation. University of Vienna.

Kelley, K.K. and King, E.G., 1961. Contributions to the Data on Theoretical Metallurgy. XIV. Entropies of the Elements and Inorganic Compounds. Bulletin 592, Bureau of Mines.

Kendall, J., Andrews, J.C., 1921. The solubilities of acids in aqueous solutions of other acids. J. Am. Chem. Soc. 43, 1545−1560.

Koch, H., Falkenburg, W.D., 1967. On the stability of chelate-complexes of polonium (IV). In: Dyrssen, D., Liljenzin, J.-O., Rydberg, J. (Eds.), Solvent Extraction Chemistry. North-Holland, Amsterdam, pp. 26−31.

Koch, H., Schmidt, H., 1963. Die bestimmung der hydrolysekonstanten in wäßrigen lösungen nach der lonenaustauschermethode. Z. Naturf., B. Chem. Biochem. Biophys., Biol. 18, 936−941.

Krestov, G.A., 1962. Thermodynamic properties of some astatine and polonium compounds. Sov. Radiochem 4, 612−617.

Kuhn, A.T., Rice, C.L., 1985. The halogens. I. Fluorine. In: Bard, A.J., Parsons, R., Jordan, J. (Eds.), Standard Potentials in Aqueous Solution. Marcel Dekker Inc, New York, pp. 67−69.

Latimer, W.M., 1952. The Oxidation States of the Elements and their Potentials in Aqueous Solutions., second ed. Prentice-Hall, Englewood Cliffs, NJ.

McCluggage, W.C., 1949. Quarterly Progress Report. Mound Laboratory Report, MLM-405-2, p. 71.

McPhail, D.C., 1995. Thermodynamic properties of aqueous tellurium species between 25 and 350 °C. Geochim. Cosmochim. Acta 59, 851−866.

Moyer, H.V., 1956. Chemical properties of polonium. In: Moyer, H.V., Gnagey, L.B., Rogers, A.J. (Eds.), Polonium. U.S. Atomic Energy Commission, pp. 33−96. , TID-5221.

Nikol'skii, B.P., Sinitsyna, G.S., Ziv, D.M., 1958. The determination of the valency of polonium in solution. Trudy Rad. Inst. Im. V. G. Khlop 8, 141−152.

Olin, Å., Noläng, B., Osadchi, E.G., Öhman, L.-O., Rosén, E., 2005. Chemical Thermodynamics of Selenium., vol. 7. Elsevier, Amsterdam.

Orban, E., 1947. Progress Report. Mound Laboratory Report, MLM-M-159.

Paneth, F., von Hevesy, F., 1913. Über die elektrochemische vertretbarkeit von radioelementen. Monatsh. 34, 1593−1603.

Panson, A.J., 1963. Polarography of the ditelluride ion. J. Phys. Chem. 67, 2177−2180.

Parsons, R., 1985. Standard electrode potentials: units, conventions and methods of determination. In: Bard, A.J., Parsons, R., Jordan, J. (Eds.), Standard Potentials in Aqueous Solution. Marcel Dekker Inc, New York, pp. 1−11.

Piontelli, R., 1966. Preface. In: Pourbaix, M. (Ed.), Atlas of Electrochemical Equilibria in Aqueous Solutions. Pergamon Press, Oxford, pp. 11−15.

Powell, K.J., Brown, P.L., Byrne, R.H., Gadja, T., Hefter, G., Leuz, A.-K., et al., 2009. Chemical speciation of environmentally significant metals with inorganic ligands. Part 3. The Pb^{2+} + OH^-, Cl^-, $CO_3{}^{2-}$, $SO_4{}^{2-}$ and $PO_4{}^{3-}$ systems. Pure Appl. Chem 81, 2425−2476.

Powell, K.J., Brown, P.L., Byrne, R.H., Gadja, T., Hefter, G., Leuz, A.-K., et al., 2011. Chemical speciation of environmentally significant metals with inorganic ligands. Part 4. The Cd^{2+} + OH^-, Cl^-, CO_3^{2-}, SO_4^{2-} and PO_4^{3-} systems. Pure Appl. Chem 83, 1163–1214.

Powell, K.J., Brown, P.L., Byrne, R.H., Gadja, T., Hefter, G., Leuz, A.-K., et al., 2013. Chemical speciation of environmentally significant metals with inorganic ligands. Part 5. The Zn^{2+} + OH^-, Cl^-, CO_3^{2-}, SO_4^{2-} and PO_4^{3-} systems. Pure Appl. Chem 85, 2249–2311.

Power, W.H., 1949a. Quarterly Progress Report. Mound Laboratory Report, MLM-368-1, p. 62.

Power, W.H., 1949b. Quarterly Progress Report. Mound Laboratory Report, MLM-379-1, p. 82.

Ringbom, A., 1953. Solubilities of Sulfides. Report to Analytical Section, IUPAC (July).

Ruzinov, L.P. and Giljanickij, B.S., 1975. Equilibrium Transformations of Metallurgical Reactions. Moscow, 1975.

Schneidt, S., 1929. Electrochemical behaviour of Po in solutions of various H-ion concentration. Sitzb. Akad. Wiss. Wien. 138, 755–765.

Shannon, R.D., 1976. Revised effective ionic radii and systematic studies of interatomic distances in halides and chalcogenides. Acta Cryst. A32, 751–767.

Sillén, L.G., Martell, A.E., 1964. Stability Constants of Metal-Ion Complexes. The Chemical Society, London, Special Publication No. 17.

Sillén, L.G., Martell, A.E., 1971. Stability Constants of Metal-Ion Complexes. The Chemical Society, London, Special Publication No. 25.

Starik, I.E., Ampelogova, N.I., 1965. Study of the complex formation of plutonium with chloride and perchlorate ions by the method of extraction. Russ. Radiochem. 7, 657–662 (Note this paper's subject is polonium even though the translated title indicates plutonium).

Starik, I.E., Ampelogova, N.I., Kuznetsov, B.S., 1964a. Hydrolysis of polonium in perchloric acid solutions. Radiokhimiya 6, 519–524.

Starik, I.E., Ampelogova, N.I., Kuznetsov, B.S., 1964b. Complex formation of polonium with the chloride ion in aqueous and aqueous-acetone solutions. Radiokhimiya 6, 524–527.

Staritzky, E., 1951. Some Compounds of Tetravalent Polonium. U.S. Atomic Energy Commission, LA-1286.

Stull, D.R., Sinke, G.C., 1956. Thermodynamic Properties of the Elements., vol. 18. American Chemical Society.

Suganuma, H., Hataye, I., 1981. Solvent extraction study on the hydrolysis of tracer concentration of Po(IV) in chloride solutions. J. Inorg. Nucl. Chem. 43, 2511–2515.

Van Muylder, J., 1966. Polonium. In: Pourbaix, M. (Ed.), Atlas of Electrochemical Equilibria in Aqueous Solutions. Pergamon Press, Oxford, pp. 572–576. , Sect. 19.5.

Younes, A., 2013. Exploration de la chimie du polonium. PhD dissertation. University of Nantes.

Younes, A., Alliot, C., Mokili, B., Deniaud, D., Montavon, G., Champion, J., 2017. Solvent extraction of polonium(IV) with tributylphosphate (TBP). Solv. Ext. Ion Exch. 35, 77–90.

Zhdanov, S.I., 1985. Sulfur, selenium, tellurium, and polonium. In: Bard, A.J., Parsons, R., Jordan, J. (Eds.), Standard Potentials in Aqueous Solution. Marcel Dekker Inc, New York, pp. 93–125.

Zikovsky, L., 1998. Precipitation and solubility of some polonium compounds, J. Radioanal. Nucl. Chem, 227. pp. 171–172.

CHAPTER 4

The pH-potential diagram for polonium

4.1 Introduction

A Pourbaix diagram for an element summarizes some of the most important features of its chemistry. Binary pH-potential diagrams represent the formulation of the equilibria of all possible thermodynamic reactions in electrochemical systems in aqueous solution, given as a function of the independent variables, measured electrode potential (E_h), and solution pH. The diagrams provide a compact pictorial summary of the electron-transfer, proton-transfer, and electron-and-proton-transfer reactions between a given element, its ions and selected solid and gaseous compounds in the presence of water (Burgers, 1966), with equilibrium conditions being represented on plane diagrams by families of straight lines (Piontelli, 1966). The diagrams are, however, limited by (1) the reactions which have been considered in establishing them and (2) the accuracy of the values assumed for the standard reduction potentials of the substances taking part in the reactions (Burgers, 1966).

Further limitations arise when a metal forms soluble complexes of great stability with other substances, for example, cyanide, then the equilibrium diagrams for the binary metal-water system have to be modified to take into account the equilibrium conditions of these complexes. This involves plotting equilibrium diagrams for a ternary system and will generally modify the domains of relative predominance of the dissolved species and the domains of thermodynamic stability (Piontelli, 1966).

A pH-potential diagram for polonium has been described by Van Muylder (1966). He sourced his data from Guillot (1931) and Gmelin (1941) and stated that the values used for the free energy of formation of three of the six substances considered in the system were very uncertain.

The thermochemical database derived for polonium in Chapter 3, Chemical thermodynamics of polonium, has been used to reconstruct the pH-potential diagram for the polonium-water system. By using values

The Aqueous Chemistry of Polonium and the Practical Application of its Thermochemistry
DOI: https://doi.org/10.1016/B978-0-12-819308-2.00004-8

based on the most recent data available, many of the uncertainties encountered by Van Muylder have been reduced, if not eliminated.

4.2 The polonium–water system

Polonium-210 is a radioactive daughter of uranium-238. Due to its relatively short half-life of 138.378 days (Kocher, 1977), it is not possible to work with gram-quantities of the element because of the intense radiation produced. Indeed Bagnall (1957) reports the difficulties he experienced while trying to obtain data using milligram amounts. Since it is therefore not possible to maintain a solution concentration of 1 mol kg^{-1} [as used by Van Muylder (1966)], a polonium concentration of 10^{-12} mol L^{-1} has been used to construct the pH-potential diagram for the polonium/water system. This is more typical of the concentrations found in "real" solutions, as evidenced elsewhere in this work. The boundaries of the domains for all respective redox couples considered for inclusion in the diagram were calculated using the Nernst equation (as outlined in Section 4.3). The data used are summarized in Table 4.1 and the diagram shown in Fig. 4.1.

4.3 Construction of pH-potential diagrams

For every electrochemical reaction involving dissolved substances, there exists an equilibrium constant whose value, for a given temperature and total pressure, is a function of the concentrations (or activities) of the reacting substances and also of a difference in electrical potential. To establish equilibrium diagrams as a function of pH and electrode potential, the influences of the pH and the electrode potential on the equilibrium characteristics of the different reactions occurring in a given system need to be studied.

The IUPAC convention has been used in this current work to define electrochemical equations, as in Eq. (4.1), written as reductions.

$$PoO_2(s) + 4H^+ + 4e^- \rightleftharpoons Po(s) + 2H_2O \tag{4.1}$$

The standard potential ($E°$) of an electrode is the potential existing when both the oxidized and reduced forms of a redox couple are in their standard states and the Nernst equation (Eq. (3.1)) can be used to calculate the electrode potential, E, at different concentrations of the redox species involved in any given process.

Table 4.1 Domains of the pH-potential diagram for the polonium–water system at zero ionic strength, 10^5 Pa and 25°C.

Couple	E° (V)	Equation
$H_2Po(aq)/Po(s)$	− 0.574	$E = E° - 0.0296 \log(H_2Po(aq)) - 0.0592\ pH$
$HPo^-/Po(s)$	− 0.624	$E = E° - 0.0296 \log(HPo^-) - 0.0296\ pH$
$Po^{2-}/Po(s)$	− 1.021	$E = E° - 0.0296 \log(Po^{2-})$
$Po(s)/Po^{2+}$	0.643	$E = E° + 0.0296 \log(Po^{2+})$
$Po(s)/H_2PoO_3(aq)$	0.827	$E = E° + 0.0148 \log(H_2PoO_3(aq)) - 0.0592\ pH$
$Po(s)/HPoO_3^-$	1.010	$E = E° + 0.0148 \log(HPoO_3^-) - 0.0740\ pH$
$Po(s)/PoO_3^{2-}$	1.211	$E = E° + 0.0148 \log(PoO_3^{2-}) - 0.0888\ pH$
Po^{2+}/PoO^{2+}	0.936	$E = E° - 0.0592\ pH$
$Po^{2+}/PoO(OH)^+$	0.961	$E = E° - 0.0888\ pH$
$Po^{2+}/H_2PoO_3(aq)$	1.010	$E = E° - 0.1184\ pH$
$PoO^{2+}/PoO_3(s)$	1.392	$E = E° - 0.0296 \log(PoO^{2+}) - 0.1184\ pH$
$PoO(OH)^+/PoO_3(s)$	1.367	$E = E° - 0.0296 \log(PoO(OH)^+) - 0.0888\ pH$
$H_2PoO_3(aq)/PoO_3(s)$	1.318	$E = E° - 0.0296 \log(H_2PoO_3(aq)) - 0.0592\ pH$
$HPoO_3^-/PoO_3(s)$	0.951	$E = E° - 0.0296 \log(HPoO_3^-) - 0.0296\ pH$
$PoO_3^{2-}/PoO_3(s)$	0.550	$E = E° - 0.0296 \log(PoO_3^{2-})$

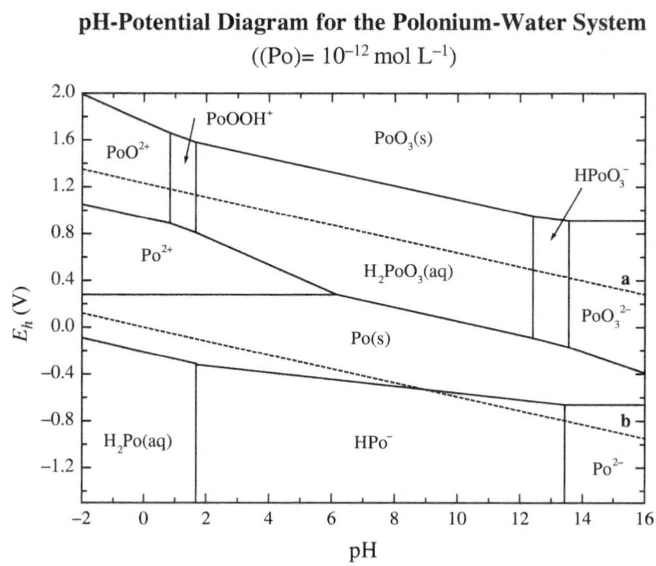

Figure 4.1 pH-potential diagram for the polonium–water system.

At 25°C, the Nernst equation may be simplified to

$$E = E^{\circ} + \frac{0.0592}{n} \log \frac{[A_{ox}]}{[A_{red}]} \tag{4.2}$$

For example, a value of 0.731 V was obtained for the E° of reaction (4.1), and hence

$$E = 0.731 + \frac{0.0592}{4} \log \frac{[PoO_2][H^+]^4}{[Po][H_2O]^2} \tag{4.3}$$

At the point where both species exist at an equal concentration, that is, the domain boundary, then

$$\begin{aligned} E &= 0.731 + \frac{0.0592}{4} 4\log\left[H^+\right] \\ &= 0.731 - 0.0592 \text{ pH} \end{aligned} \tag{4.4}$$

The reversible electrode potentials for hydrogen and oxygen are also considered when constructing a pH-potential diagram and their derivation follows (Potter, 1956). The potential of the reversible hydrogen electrode under any given conditions of hydrogen gas pressure and hydrogen ion concentration can be calculated by applying the Nernst equation to the overall electrode equilibrium,

$$2H^+ + 2e^- \rightleftharpoons H_2 \tag{4.5}$$

viz

$$E = E^{\circ} + \frac{2.303RT}{2F} \log \frac{[A_{H^+}]^2}{[A_{H_2}]} \tag{4.6}$$

When the activities of the hydrogen ion and hydrogen gas are both unity, then $E = E^{\circ}$. Under these conditions, it is the convention that $E = 0$ and hence $E^{\circ} = 0$, and therefore, Eq. (4.6) becomes

$$E = \frac{2.303RT}{2F} \log \frac{[A_{H^+}]^2}{[A_{H_2}]} \tag{4.7}$$

At 25°C, Eq. (4.7) becomes

$$E = 0.0592 \log A_{H^+} - 0.0296 \log A_{H_2} \tag{4.8}$$

At standard pressure, Eq. (4.8) reduces to

$$E = 0.0592 \log A_{H^+} \tag{4.9}$$

Eq. (4.9) can also be expressed in terms of the hydrogen ion component, pH, so that

$$E = -0.0592\,\mathrm{pH} \tag{4.10}$$

This line, denoted "**a**," is shown in Fig. 4.1 and represents the reduction equilibrium of water.

In a similar manner, the potential of the reversible oxygen electrode can be calculated from the overall electrode equilibrium,

$$O_2 + 4H^+ + 4e^- \rightleftharpoons 2H_2O \tag{4.11}$$

viz

$$E = E^{o} + \frac{2.303\mathrm{RT}}{4\mathrm{F}} \log \frac{[A_{O_2}][A_{H^+}]^4}{[A_{H_2O}]^2} \tag{4.12}$$

The E^{o} for the reaction is 1.229 V (Sillén and Martell, 1964; Hoare, 1985; Brown and Ekberg, 2016). When the activities of oxygen gas and water are unity, Eq. (4.12) becomes

$$E = 1.229 + \frac{2.303\mathrm{RT}}{4\mathrm{F}} \log\,[A_{H^+}]^4 \tag{4.13}$$

At 25°C, Eq. (4.13) becomes

$$E = 1.229 + 0.0592\ \log\,[A_{H^+}] \tag{4.14}$$

In terms of the hydrogen ion component, pH, Eq. (4.14) reduces to

$$E = 1.229 - 0.0592\,\mathrm{pH} \tag{4.15}$$

This line, denoted "**b**," is shown in Fig. 4.1 and represents the oxidation equilibrium of water.

These two lines, **a** and **b**, in Fig. 4.1 identify the region between which water is thermodynamically stable, that is, the electrolysis of water into hydrogen or oxygen does not occur.

For electrode reactions occurring on a pH-potential diagram, the reversible electrode potential of, for example, polonium metal in equilibrium with Po^{2+} ions is given by Eq. (4.16)

$$Po^{2+} + 2e^- \rightleftharpoons Po(s) \tag{4.16}$$

The equilibrium is seen to be independent of pH and is, therefore, expressed as a horizontal line corresponding to a potential of 0.643 V.

For equilibrium reactions depending only on pH and not potential (i.e., pure proton-transfer reactions not involving electrons), for example,

$$HPoO_3^- \rightleftharpoons PoO_3^{2-} + H^+ \tag{4.17}$$

the equilibrium is shown as a vertical line parallel to the potential axis.

Lines representing the conditions under which the activities of two dissolved substances are equal, for example, $HPoO_3^-$ and PoO_3^{2-} (pH = 13.55), correspond to *the domains of relative predominance* of these dissolved forms. Either side of this boundary, *both* species will be present, although the relative proportions of each will alter as the distance from the boundary increases. This is also true for equilibrium conditions between a solid and a dissolved substance, for example, Po(s) and H_2PoO_3(aq).

Lines representing the equilibrium conditions between two solid substances, for example, PoO_2(s) and Po(s) in the following equation

$$PoO_2(s) + 4H^+ + 4e^- \rightleftharpoons Po(s) + 2H_2O \tag{4.18}$$

correspond to *the domains of relative stability* of these two solids. The boundary denotes equilibrium between the two solids. However, on either side of the boundary, only *one* solid species can exist, if a solid can be formed.

References

Bagnall, K.W., 1957. Chemistry of the Rare Radioelements. Butterworths Scientific Publications, London.

Brown, P.L., Ekberg, C., 2016. Hydrolysis of Metal Ions. Wiley-VCH, Weinheim.

Burgers, W.G., 1966. Foreword. In: Pourbaix, M. (Ed.), Atlas of Electrochemical Equilibria in Aqueous Solutions. Pergamon Press, Oxford.

Gmelin, L., 1941. Gmelins Handbuch der Anorganischen Chemie: Polonium. Verlag Chemie, Berlin, S.N. 13.

Guillot, M., 1931. Sur les conditions de précipitation du polonium et sur quelques-uns de ses dérivés complexes. J. Chim. Phys. 28, 14–41.

Hoare, J.P., 1985. Oxygen. In: Bard, A.J., Parsons, R., Jordan, J. (Eds.), Standard Potentials in Aqueous Solution. Marcel Dekker Inc, New York, pp. 49–66.

Kocher, D.C., 1977. Nuclear Decay Data for Radionuclides Occurring in Routine Releases From Nuclear Fuel Cycle Facilities. Oak Ridge National Laboratory, ORNL/NUREG/T102.

Piontelli, R., 1966. Preface. In: Pourbaix, M. (Ed.), Atlas of Electrochemical Equilibria in Aqueous Solutions. Pergamon Press, Oxford, pp. 11–15.

Potter, E.C., 1956. Electrochemistry—Principles and Applications. Cleaver-Hume Press Ltd, London.

Sillén, L.G., Martell, A.E., 1964. Stability Constants of Metal-Ion Complexes. The Chemical Society, London, Special Publication No. 17.

Van Muylder, J., 1966. Polonium. In: Pourbaix, M. (Ed.), Atlas of Electrochemical Equilibria in Aqueous Solutions. Pergamon Press, Oxford, pp. 572–576. Section 19.5.

CHAPTER 5

The use of pH–potential diagrams in practical applications

5.1 Introduction

The pH–potential diagram provides a compact, pictorial summary of electron-transfer, proton-transfer, and electron-and-proton-transfer reactions, favored on thermodynamic grounds, for any given system. Such diagrams provide a basic framework for understanding solution chemistry and can be applied across a wide range of studies, from simple metal-water systems to the more complicated systems of many industrial processes. They must, however, be used with caution. Although thermodynamics determines the direction in which an overall reaction will tend, the rate of reaction will depend on the kinetics. Also, the values of pH in the diagram always refer to the solution in the immediate vicinity of an electrode. The local pH near the electrode, however, can vary from the bulk pH of the solution if the electronation reaction taking place at the electrode consumes hydrogen ions or generates hydroxyl ions. For example, the pH-potential diagram may indicate that a particular hydroxide will only form above a certain pH value, while experimentally it is observed that the hydroxide forms at a much lower pH (Bockris and Reddy, 1970).

This chapter describes several case studies to illustrate how these diagrams can be used in conjunction with practical observations and measurements to gain a better understanding of the aqueous chemistry of polonium.

5.2 Derivation of pH–potential diagrams

The pH–potential diagrams for the following systems have been constructed in a similar manner to that of the polonium–water system (as outlined in Chapter 4: The pH–potential diagram for polonium):

1. lead–water system;
2. selenium–water system;
3. tellurium–water system;

DOI: https://doi.org/10.1016/B978-0-12-819308-2.00005-X

4. polonium—sulfur—water system;
5. lead—sulfur—water system;
6. polonium—cyanide—water system;
7. lead—cyanide—water system;
8. polonium—chlorine—water system;
9. lead—chlorine—water system;
10. polonium—nitrogen—water system; and
11. lead—nitrogen—water system.

Again, the boundaries of the domains for all respective redox couples considered for inclusion in the diagrams were calculated using the usual method. The data used to derive the diagrams are summarized in Chapter 3, Chemical thermodynamics of polonium or Conclusion. The respective diagrams appear as figures within the relevant case study sections. Where necessary, operating domains for the processes described have been highlighted on the diagrams as boxes, which describe the range of values likely to be found.

5.3 The aqueous speciation of polonium, selenium, tellurium, and lead

Selenium and tellurium occur as impurities in many sulfide ores, especially those with copper pyrites. The chief commercial source of both elements is anode slimes from the electrolytic recovery of copper (Kolthoff and Elving, 1961a). The most important lead ore is galena (PbS), with other less important minerals being cerrusite ($PbCO_3$) and anglesite ($PbSO_4$). Lead also occurs in nature in association with other metals, notably silver and zinc (Kolthoff and Elving, 1964).

The pH—potential diagrams for the selenium—, tellurium—, and lead—water systems are illustrated in Figs. 5.1, 5.2, and 5.3, respectively. In a similar manner to that of the polonium—water diagram, these diagrams have been constructed using "typical" solution concentrations, in this case, those that may be within the range encountered during anode slimes processing, which is discussed in detail in Section 5.4.

Like its Group VI homologue polonium (see Fig. 4.1), within the domain of the thermodynamic stability of water, selenium occurs predominantly as the element at low E_h and as aqueous species in acidic, neutral, and alkaline regions at higher E_h. Similar to polonium, selenium also exhibits a region where selenide ions are stable in the domain where water is thermodynamically stable (high pH and low E_h). In contrast at

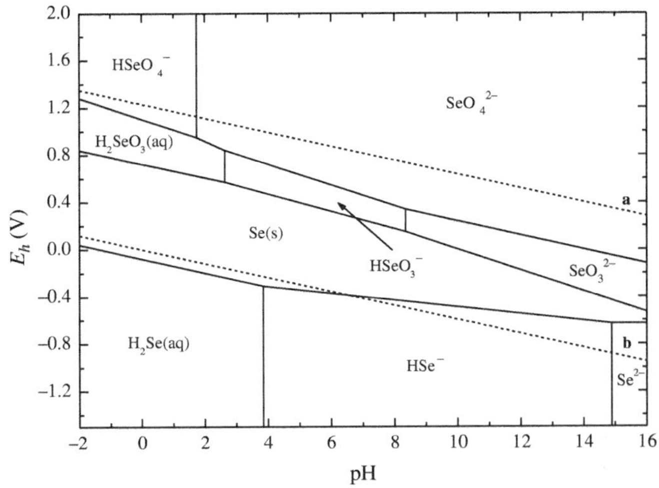

Figure 5.1 pH–potential diagram for the selenium–water system [(Se) = 0.16 mol L^{-1}].

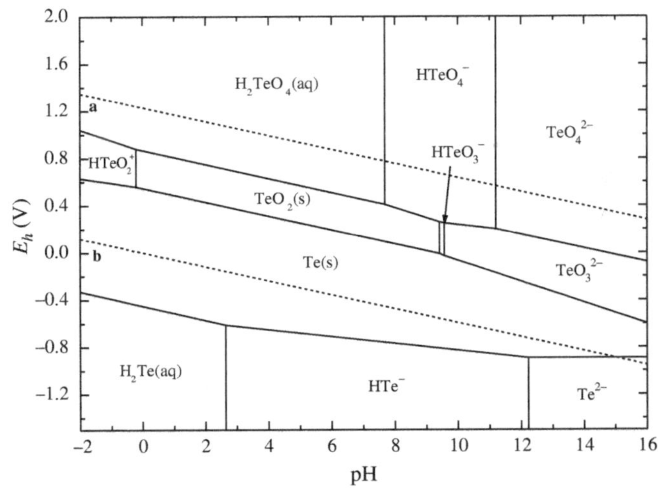

Figure 5.2 pH–potential diagram for the tellurium–water system [(Te) = 0.020 mol L^{-1}].

higher E_h, solid tellurium oxide (TeO_2(s)) predominates over a substantial region (pH range −0.5 to 9) whereas solid lead oxide (PbO(s)) is the predominant lead phase in the pH range 8.5–13.5. Above pH 13.5, Pb $(OH)_3^-$ is the dominant species.

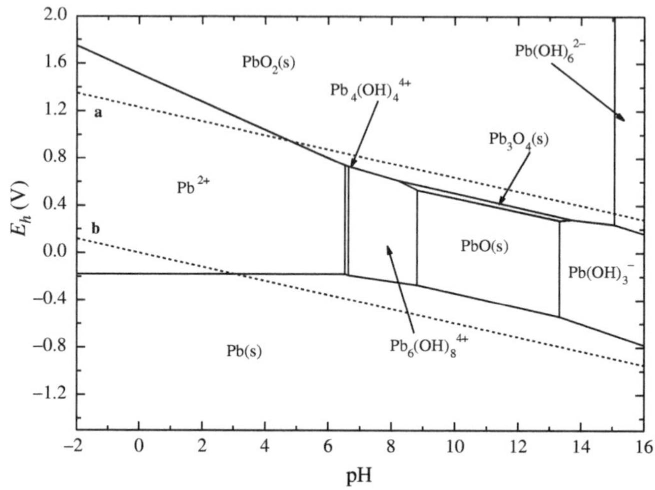

Figure 5.3 pH–potential diagram for the lead–water system [(Pb) = 0.020 mol L^{-1}].

5.4 Case study—polonium behavior during anode slimes processing

Copper is present in the Earth's crust mainly in the form of sulfide minerals such as chalcopyrite ($CuFeS_2$), bornite (Cu_5FeS_4), and chalcocite (Cu_2S). It also occurs to a much lesser extent in the form of oxidized minerals (carbonates, oxides, silicates, sulfates) (Biswas and Davenport, 1994). Most copper is processed using a combination of mining, concentrating, smelting, and refining techniques. Copper ore may be obtained from either underground or surface mines and rarely contains a sufficiently high percentage of copper to allow direct smelting. Ores containing copper sulfide minerals are crushed and ground and the copper minerals recovered using flotation to produce concentrates containing from 25 to 55 wt.% copper. Copper is then extracted by smelting using heat, flux, and oxygen in a furnace to produce matte or blister copper which is then fire-refined and cast into copper anodes. These anodes are electrolytically refined in an acidic copper sulfate electrolyte, with the copper being deposited on starting sheets to produce a cathode product. Anode impurities either dissolve in the electrolyte or fall to the bottom of the electrolytic cell as anode slimes. These slimes often contain selenium, tellurium, gold, silver, and platinum group metals (PGMs) and can represent a very significant value. The recovery of by-products from the anode slimes is, therefore, an important operation (Kroschwitz, 1993).

Table 5.1 Potential fractions of anode elements entering slimes.

Metal	% to slimes
Cu	<0.2
Au	99
Ag	98
Se	98
Te	98
Pb	98
Sb	50
As	30
Co	5
Ni	5
Fe	0
Zn	0
Bi	a

[a]Dissolves up to 0.2 g L^{-1} in electrolyte and then forms slimes.

Sulfur, selenium, and tellurium in the slimes can combine with copper and silver to form insoluble sulfides, selenides, and tellurides, respectively. Gold most frequently exists as the metal and in combination with tellurium, while the PGMs are typically associated with the sulfides. Arsenic, bismuth, and antimony can enter the slimes, depending on their concentrations in the electrolyte. Lead, tin, nickel, and cobalt also precipitate. Typical fractions of anode elements entering the slimes are listed in Table 5.1 (Biswas and Davenport, 1994).

In some deposits, uranium-bearing minerals such as uraninite (UO_2), coffinite [$U(SiO_4)_{1-x}(OH)_{4x}$ ($x = 0-1$)], and brannerite ($U(Ti,Fe)_2O_6$) are associated with the copper mineralogy (Vonk, 1993). Polonium-210, produced from the decay of ^{238}U, is the longest lived naturally occurring isotope of polonium [$t_{1/2} = 138.378$ days (Kocher, 1977)]. Ores containing 500 mg kg^{-1} of uranium have about 100 μg of polonium per tonne of ore (Ansoborlo et al., 2012). When copper ores containing uranium minerals are processed, some ^{210}Po is found in the anode slimes, whereas the uranium parent is removed in the slag phase during smelting.

During anode slimes treatment, polonium, depending on deportment during processing, can potentially result in unacceptably high levels of radioactive contamination in final products. Although techniques are available for polonium removal during treatment, none are completely effective.

To ensure that specifications are met in final products, it is necessary to identify and closely monitor where the impurities, including ^{210}Po, partition during anode slimes processing. Although polonium can be measured in both solid and liquor streams using radioanalytical techniques, the results give no indication of its likely chemical forms. By studying both the chemistry and mineralogy of polonium and its neighboring elements in the periodic table (lead, bismuth, selenium, and tellurium), it may be possible to make some conclusions regarding the aqueous speciation and solid phase siting of ^{210}Po.

5.4.1 Anode slimes processing

The anode slimes collected from the bottom of electrolytic cells are processed to recover copper, selenium, tellurium, and precious metals. In general, refineries employ a combination of both pyrometallurgical and hydrometallurgical techniques (Brown, 2001).

Decopperization is the first step in most anode slimes treatment plants. Copper may be removed by leaching in sulfuric acid at atmospheric pressure, sulfuric acid pressure leaching, high temperature oxidizing roast, or sulfatizing roast with sulfuric acid followed by either water or acid leaching. Selenium is removed by smelting with sodium carbonate (Na_2CO_3) and sodium nitrate ($NaNO_3$), roasting with sodium carbonate or roasting with sulfuric acid. Generally, a combination of techniques is used depending on other impurities contained in the slimes. Tellurium removal is dependent on its concentration and also the mineralogy of the slimes and is generally leached along with copper during hydrometallurgical treatment.

The solids remaining after impurity removal are smelted to produce a doré, Au/Ag, metal which is then cast into anodes ready for precious metals refining. A typical analysis of doré metal is contained in Table 5.2 (Kroschwitz, 1993).

Table 5.2 Typical doré metal assay.

Metal	wt.%
Au	8–9
Ag	86–92
Cu	0.5–1.0
Pd	0.16–0.18
Pt	0.005–0.009
Pb	0.02
Te	0.003
Se	0.00002

Alternatively, impurities may be directed to the solid phase by employing similar technology to that used in the refining of gold ores (Kroschwitz, 1994). After decopperization, gold, silver, and PGMs are leached using cyanide solution. The precious metals are then recovered using zinc or aluminum dust and, after acid treatment to remove excess precipitant, smelted to produce doré metal.

A typical process for treatment of doré metal is as follows (Leigh, 1981). The silver is first separated by electrolysis and the cathode then melted and cast into silver ingots. The silver anode slimes are leached with acid to remove any remaining silver, and then cast into anodes and electrorefined to recover the gold. The gold cathodes are melted and cast into ingots. Platinum and palladium accumulate in the gold chloride electrolyte and are precipitated as ammonium chloroplatinate and chloropalladate, respectively.

Hydrometallurgical treatment of copper anode slimes leads to the generation of solid and liquor streams in various process steps. A typical flowchart for these processes, that is used at Olympic Dam in South Australia, is illustrated in Fig. 5.4 (Hall, 1993).

As shown in Fig. 5.4, anode slimes are initially leached using sulfuric acid, with air- and steam-sparging at atmospheric pressure, to produce a decopperized cake. The concentration of sulfuric acid used depends on the ore being processed, but is typically within the range of 40–180 g L^{-1} (Amer, 2002; Saeedi et al., 2013; Dimitrijevic et al., 2014; He et al., 2014; Lu et al., 2015). Analysis of these sulfuric acid solutions in the presence of a typical anode slimes indicates that the pH would range from −0.3 to 0.3 with a potential between about 600 and 700 mV. The decopperized cake can then be treated with sodium cyanide to dissolve gold, silver, and PGMs (cyanidation). Precious metals recovery can be carried out by precipitation using zinc dust (zinc precipitation).

The typical range in E_h–pH conditions for cyanidation and zinc precipitation have been indicated by Marsden and House (2006) (see Fig. 5.5). The excess zinc used in the precipitation can be removed using sulfuric acid with E_h–pH conditions similar to those used in decopperization (aciding). Finally, a mixture of hydrogen peroxide and sodium nitrate can be added to the remaining solid. Under these conditions, selenium and tellurium are oxidized and silver is dissolved in the nitric acid that forms. The silver can then be recovered using precipitation by adding sodium chloride (Hall, 1993).

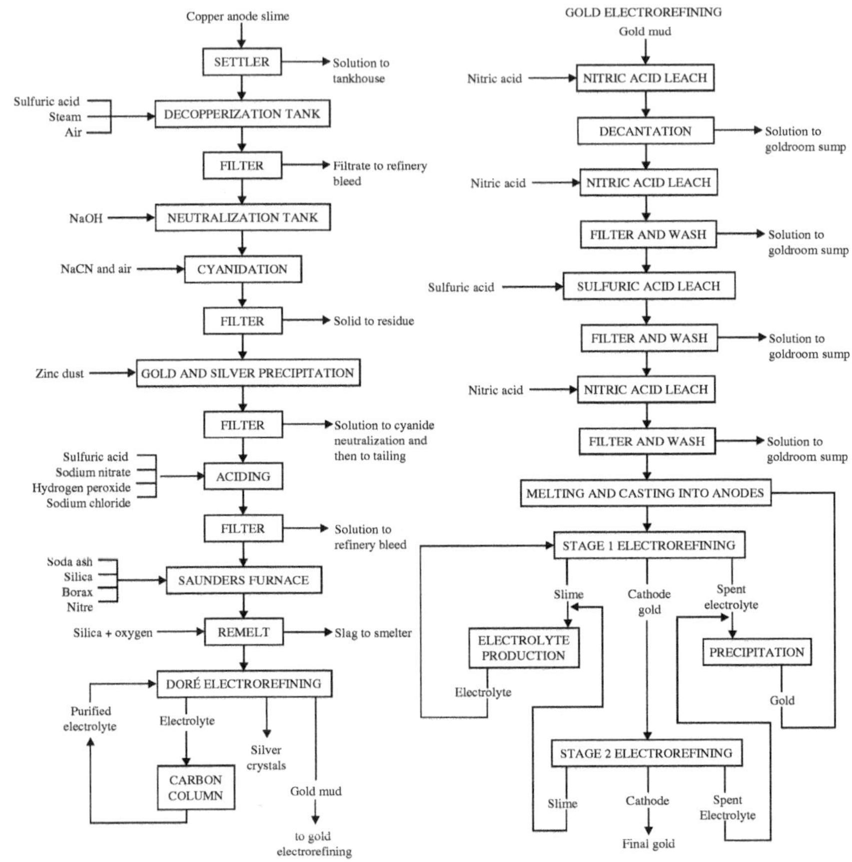

Figure 5.4 Anode slimes processing at Olympic Dam, South Australia. *Adapted from Hall, S. 1993. Gold and silver recovery from copper anode slimes at The Olympic Dam Joint Venture, Roxby Downs, SA. In: Woodcock, J.T., Hamilton, J.K. (Eds.), Australasian Mining and Metallurgy — The Sir Maurice Mawby Memorial Volume, second ed., vol. 2. The Australasian Institute of Mining and Metallurgy, Parkville, pp. 1102–1105.*

The typical deportment of polonium, lead, bismuth, selenium, and tellurium in the various process steps is indicated in Table 5.3. Further, the major solid phases that may be present in each process are indicated in Table 5.4 (Brown, 1998).

5.4.1.1 Copper anode (raw) slimes

The general behavior of copper anode impurities during electrorefining in acidic copper sulfate has been described previously (Kroschwitz, 1993, 1995). Gold, silver, and PGMs do not dissolve in the electrolyte, and thus

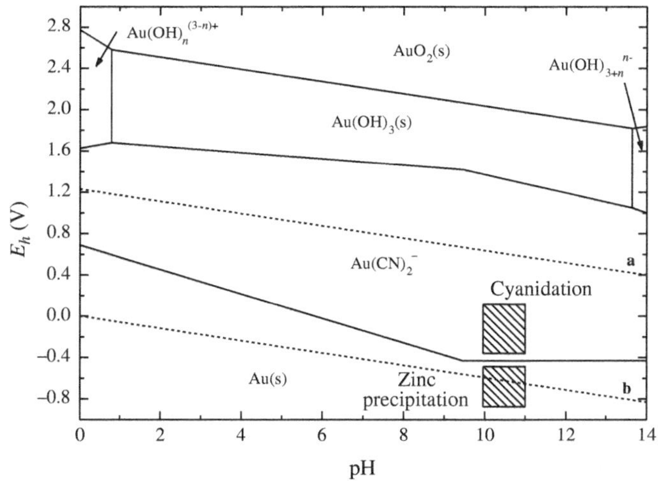

Figure 5.5 Indicative E_h–pH ranges for the cyanidation and zinc precipitation processes on the pH–potential diagram for the gold–water–cyanide system. *Adapted from data given by Marsden, J.O., House, C.I., 2006. The Chemistry of gold Extraction. Society for Mining, Metallurgy and Exploration, second ed. Littleton and Yannopoulos, J.C., 1991. The Extractive Metallurgy of Gold. Van Nostrand Reinhold, New York; the species $Au(OH)_n^{(3-n)+}$ represents cationic gold-hydroxide complexes with $n = 0–2$ and $Au(OH)_{3+n}^{n-}$ anionic species with $n = 1$ or 2.*

Table 5.3 Analytical results for anode slimes solids and liquors (% of feed).

Process	Stream	Pb	Bi	Se	Te	Po
Decopperization	Liquor	0.1	24	<0.5	<0.5	<0.1
	Solid	99.9	76	100	100	100
Cyanidation	Liquor	0.5	<0.5	100	100	0.6
	Solid	99.5	100	<0.1	<0.1	99.4
Zinc precipitation	Liquor	<0.5	<0.5	65	6	<0.5
	Solid	100	100	35	94	100
Aciding and silver recovery	Liquor	3	<0.5	98	95	6
		97	100	2	5	94

are present in the slimes as free or bound metal phases. Selenium and tellurium, present in the anodes as selenide phases (e.g., Ag_2Se, Cu_2Se) and telluride phases (e.g., Ag_2Te_4), enter the slimes in these bound forms. Metals less noble than copper, for example, nickel, iron, and lead, dissolve from the anode. Lead then forms a sulfate precipitate, which is insoluble in the electrolyte, and falls into the slimes. Bismuth, also less noble than

Table 5.4 Major phases in the solids of the processing steps of anode slimes processing (elements in bold represent major constituents of multielement phases).

Solid	Phase	Formula
Raw slimes	Te-bearing eucairite	(**Ag**, **Au**, **Cu**, Bi, Pb)$_2$(**Se**, Te)
	Ag-bearing umangite	(**Cu**, **Ag**)$_3$(**Se**, Te)$_2$
	Geminite	$Cu_2As_2O_7$
	Copper arsenate	(**Cu**, Ag)$_7$(**As**, Te, S)$_3$O$_{12}$
	Rooseveltite	(**Bi**, **Pb**, Sb, Cu)AsO$_4$
Decopperization residue	Cu-leached eucairite	(**Ag**, **Au**, Cu, Bi, Pb)$_2$(**Se**, Te)
	Cu-leached umangite	(**Ag**, Cu)$_3$(**Se**, Te)$_2$
	Anglesite	$PbSO_4$
Cyanidation residue	Lead oxide hydrate	$PbO \cdot H_2O$
	Secondary naumannite	(**Ag**, **Au**, Cu, Pb)$_2$(**Se**, Te)
Zinc precipitate	Au-bearing naumannite	(**Ag**, **Au**, Pb)$_2$(**Se**, Te)
	Fischesserite	Ag_3AuSe_2
	Wulfingite	$Zn(OH)_2$
	Zincite	ZnO
Silver chloride solid	Chloroargyrite	AgCl
	Gold	Au

copper, dissolves up to about 0.15–0.2 $g\,L^{-1}$ in the electrolyte, the remainder reporting to the slimes. Polonium-210 is also found in the raw slimes when copper ores containing uranium minerals are processed. Its mineralogy, however, is not known.

Mineralogical analyses of raw slimes (Brown, 1998) have also shown that selenium and tellurium can be present as Ag-bearing umangite [(**Cu**, **Ag**)$_3$(**Se**,Te)$_2$] and Te-bearing eucairite [(**Ag**,**Au**,**Cu**,Bi,Pb)$_2$(**Se**,Te)]; the elements in bold representing major constituents of multielement phases. The major phase containing lead is anglesite ($PbSO_4$). Lead-bearing rooseveltite [(**Bi**,**Pb**,Sb,Cu)AsO$_4$], an arsenate, may also be present although it is likely to be sparse. Copper may be found as basic sulfate minerals. Typically, no distinct telluride phases are present in raw slimes.

5.4.1.2 Decopperization

Raw anode slimes are leached using sulfuric acid to produce a decopperized cake. This leaching process dissolves phases containing arsenate as well as any basic copper sulfate minerals. In addition, copper is selectively dissolved from the selenide, which may contain telluride, phases. Phases containing lead will be solubilized, but lead will be found in the

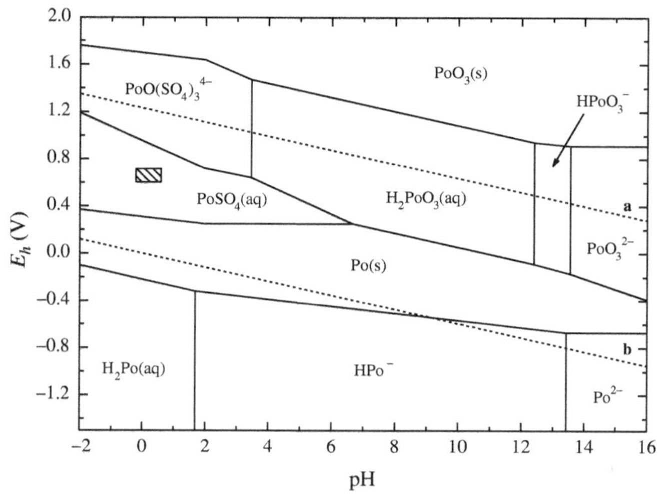

Figure 5.6 pH–potential diagram for the polonium–sulfur–water system [(Po) = 2×10^{-12} mol L^{-1}; (S) = 1.04 mol L^{-1}].

decopperized residue due to precipitation of anglesite (which may contain some bismuth). Table 5.3 indicates that typically lead, selenium, tellurium, and polonium will not be solubilized, but remain in the decopperized cake, whereas a significant proportion of bismuth might be expected to remain in the sulfuric acid liquor.

A total polonium concentration (solid plus liquor) in the decopperization circuit of 2×10^{-12} mol L^{-1} has been utilized and a pH–potential diagram for the polonium–sulfate–water system (Fig. 5.6) has been derived for this concentration. (In Sections 5.4 and 5.5, concentration describes the total concentration of an element in both the solid and liquor phases of a particular process.) Under typical operating conditions [as indicated earlier, a pH of −0.3 to 0.3 and an E_h of 600–700 mV; as indicated by the hashed box in Fig. 5.6 (and subsequent figures)], polonium should exist as the aqueous species, $PoSO_4$(aq), although the results shown in Table 5.3 suggest that polonium has solid phase speciation during decopperization. Possible reasons for the nonleachability of polonium during decopperization are: (1) siting within matrices not amenable to leaching; (2) dissolution and then coprecipitation within a solid phase not amenable to leaching (e.g., polonium may coprecipitate with anglesite, see Section 5.4.5); and/or (3) kinetic limitations.

The pH–potential diagram for lead in the decopperization circuit (Fig. 5.7), derived for a concentration of 0.020 mol L^{-1}, indicates that, under the range of operating conditions, the element will be present as $PbSO_4$(s) in the decopperized slimes. This solid phase speciation is confirmed by the assay and mineralogical results given in Tables 5.3 and 5.4.

The pH–potential diagrams for the silver–selenium– and silver–tellurium–water systems (Figs. 5.8 and 5.9, respectively), have been

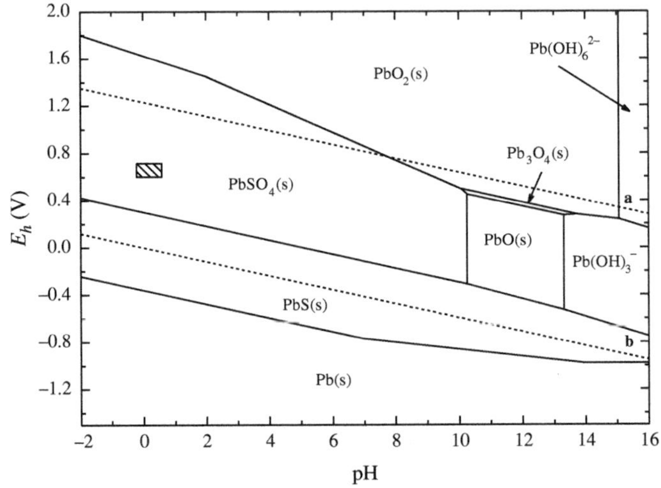

Figure 5.7 pH–potential diagram for the lead–sulfur–water system [(Pb) = 0.020 mol L^{-1}; (S) = 1.04 mol L^{-1}].

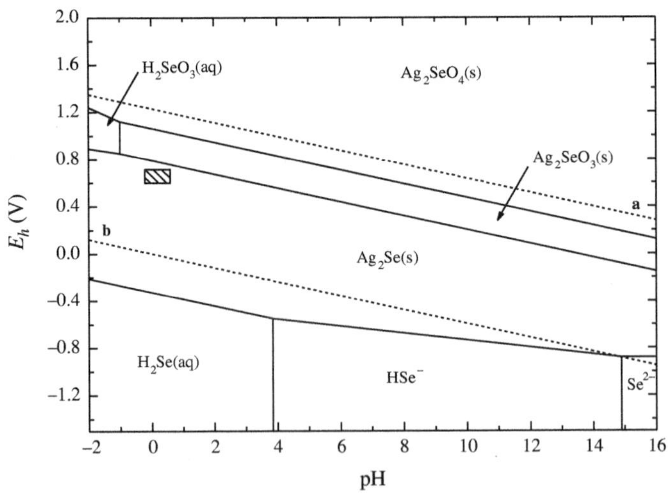

Figure 5.8 pH–potential diagram for the silver–selenium–water system [(Se) = 0.19 mol L^{-1}; (Ag) = 0.13 mol L^{-1}].

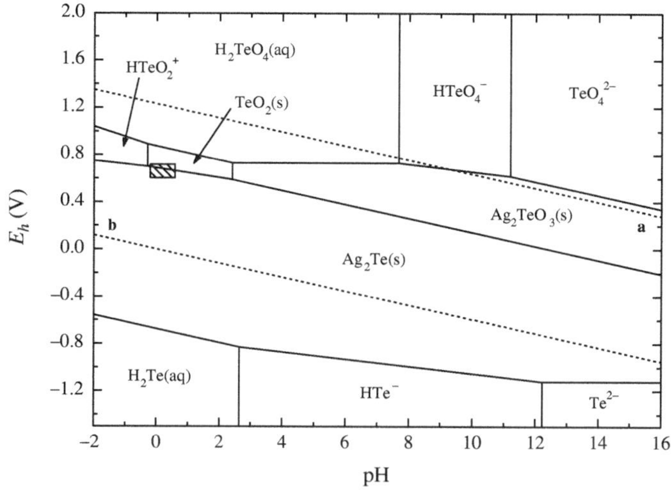

Figure 5.9 pH–potential diagram for the silver–tellurium–water system [(Te) = 0.024 mol L^{-1}; (Ag) = 0.13 mol L^{-1}].

derived for a silver concentration of 0.13 mol L^{-1} and selenium and tellurium concentrations of 0.19 and 0.024 mol L^{-1}, respectively, as found in decopperization. The diagrams indicate that both elements do not dissolve under the operating conditions. Selenium will be present as Ag_2Se(s) and the results in Tables 5.3 and 5.4 support this solid phase speciation. Fig. 5.9 indicates that tellurium could be present as either Ag_2Te(s) or TeO_2(s); however, the mineralogical analysis suggests the speciation should be as the solid telluride. It is probable that tellurium occurs as Ag_2Te(s), being present in a solid solution with Ag_2Se(s). Thus the speciation indicates that polonium, lead, selenium, and tellurium remain within the residue during decopperization, as confirmed by the data listed in Table 5.4.

5.4.1.3 Cyanidation

Sodium cyanide is used to treat the decopperized cake to dissolve gold, silver, and the PGMs. Cyanidation will destroy all of the primary selenides/tellurides, leaving a lead oxide residue containing bismuth [(**Pb**,Bi)O(s)]; minium Pb_3O_4(s), was also found. The amount of bismuth in the oxide will be similar to that associated with anglesite produced during the decopperization process. However, some gold and silver can remain in the solids during cyanidation, present in, for example, secondary naumannite (**Ag**,**Au**,Cu,Pb)$_2$(Se,Te). The phase is rare but, typically has a

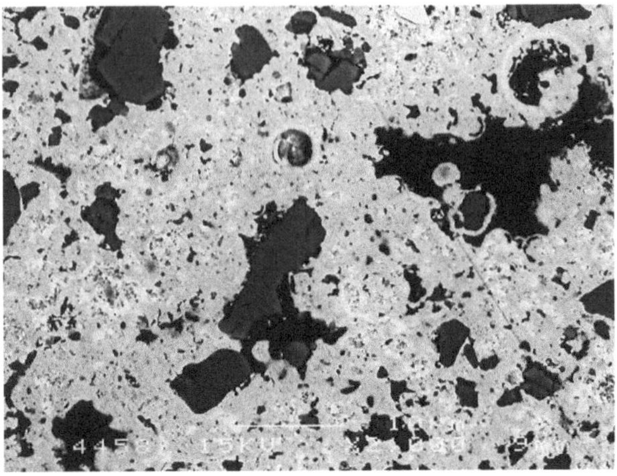

Figure 5.10 SEM backscattered electron micrograph of anode slimes residue after cyanidation (Detail of fragment rich in massive secondary naumannite (gray areas) surrounding barite and lead oxide particles (black areas) × 2000).

well-defined stoichiometry indicative of the Ag_2Se phase (Fig. 5.10). As indicated in the figure, naumannite has a "massive" microstructure, which completely encloses the barite and lead oxide particles, strongly suggesting that the phase is a secondary precipitate. All of the selenium and tellurium, less than 1% of lead and polonium and none of the remaining bismuth were found in the cyanide solution.

The solubility of a metal can be enhanced by complexation using a suitable ligand. In coordination chemistry, the cyanide ion is considered to be one of the strongest "soft" ligands, forming very stable complexes with "soft" metals such as gold and silver. Gold and silver are leached from the decopperized slimes by oxidation of the metals using air, the process forming the stable dicyanoaurate and dicyanoargentate complexes, respectively. The reactions can be described by Eq. (5.1).

$$Au + Ag + 4CN^- + \frac{1}{2}O_2 + H_2O \rightleftharpoons Au(CN)_2^- + Ag(CN)_2^- + 2OH^- \tag{5.1}$$

Polonium was not detected to any significant extent in the cyanidation liquor and this indicates that the phases in the decopperized slimes containing polonium are insoluble and remain with the cyanided cake. The pH–potential diagram for polonium in cyanide media at a concentration of 2×10^{-12} mol L^{-1} (Fig. 5.11) which indicates, under the operating

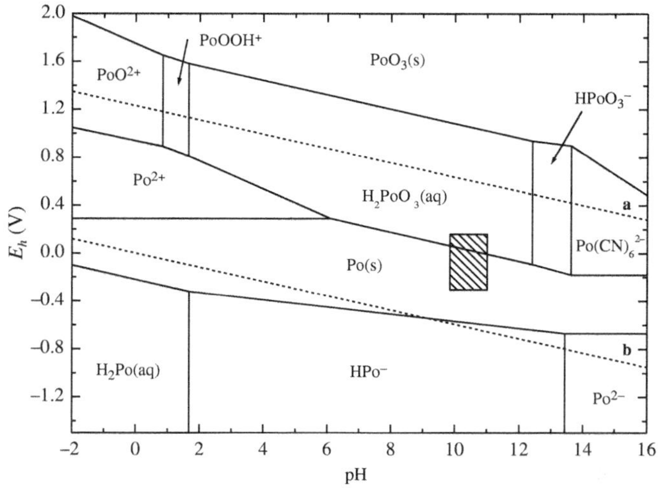

Figure 5.11 pH–potential diagram for the polonium–cyanide–water system [(Po) = 2×10^{-12} mol L^{-1}; (CN) = 0.71 mol L^{-1}].

conditions of cyanidation (see Fig. 5.5), the element will be present predominantly as Po(s). There is a smaller region of stability where polonium could be present as the soluble species, H_2PoO_3(aq), but in both cases [whether it is predicted to be present as Po(s) or H_2PoO_3(aq)], the polonium is likely to remain with lead in the PbO(s) phase [it was indicated that polonium will likely be associated with $PbSO_4$(s) during decopperization and, as such, it might be expected to remain with lead during cyanidation]. However, if polonium did become soluble, the diagram also indicates how control of the conditions could be used to direct polonium into the elemental phase, for example, by decreasing the E_h.

Under operating conditions for cyanidation (i.e., high pH), lead forms weak complexes with CN^- relative to the stability of lead-hydroxide complexes (Sillén and Martell, 1964) and will, therefore, not dissolve. The pH–potential diagram for lead in cyanide media for a concentration of 0.02 mol L^{-1}, as found in cyanidation (Fig. 5.12), indicates that the element will be present as PbO(s) in the cyanided solid, and this solid speciation is confirmed by the data listed in Tables 5.3 and 5.4. The diagram also indicates that if the pH became too high during cyanidation, lead would likely become solubilized, and the likely species formed would be $Pb(CN)_4^{2-}$.

Selenium and tellurium behavior during cyanidation has not been well documented. However, complexation reactions are known to occur

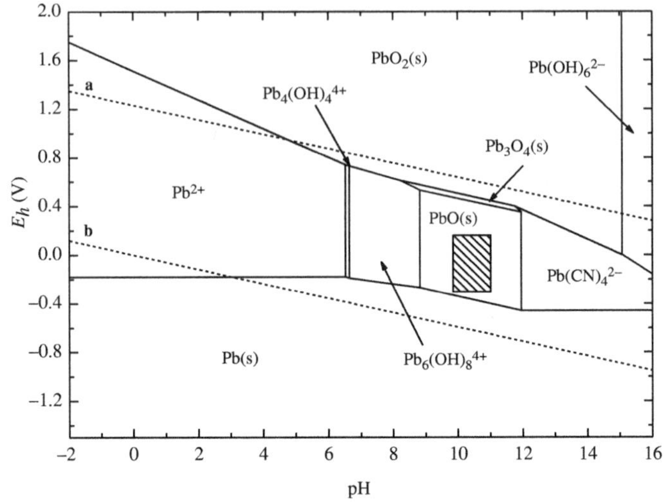

Figure 5.12 pH−potential diagram for the lead−cyanide−water system [(Pb) = 0.02 mol L^{-1}; (CN) = 0.71 mol L^{-1}].

between CN^{-} and either sulfur or selenium. Kolthoff and Elving (1961b) and Kroschwitz (1997) describe the formation of alkali selenocyanates from the reaction of selenium and alkali cyanides and Kolthoff and Elving (1961b) indicate that tellurium is only slightly soluble in such solutions. If selenium does form a selenocyanide complex, it is feasible that gold and silver could react with it to form metal selenocyanate complexes similar to those described for cyanide by reaction (5.1); however, no confirmation of the formation of such species has been published for the process conditions used in cyanidation.

Incomplete data are available to construct and study the pH−potential diagrams for the selenium− and tellurium−cyanide−water systems, due to the lack of necessary information for selenocyanide and tellurocyanide species. Consequently, the selenium− and tellurium−water diagrams (Figs. 5.13 and 5.14), derived for concentrations of 0.12 and 0.019 mol L^{-1}, respectively, as found in cyanidation, were used to indicate the behavior of both elements. The diagrams show that the likely speciation will be either as the aqueous anionic species SeO_3^{2-} and TeO_3^{2-} or the elemental phases Se(s) and Te(s). The results given in Table 5.3 suggest that both species should only be present in the aqueous phase, implying that either the E_h needs to be maintained to the upper region of the range indicated or that the formation of soluble seleno- (and possibly telluro-) cyanide species is possible, therefore promoting the stability of aqueous species to lower E_h.

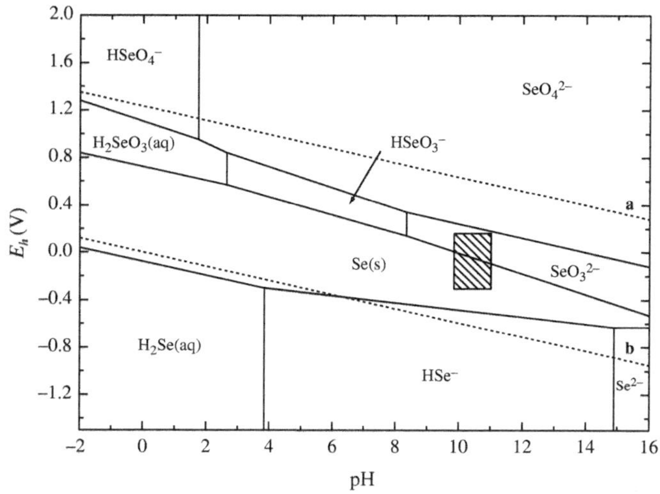

Figure 5.13 pH–potential diagram for the selenium–water system [(Se) = 0.12 mol L^{-1}].

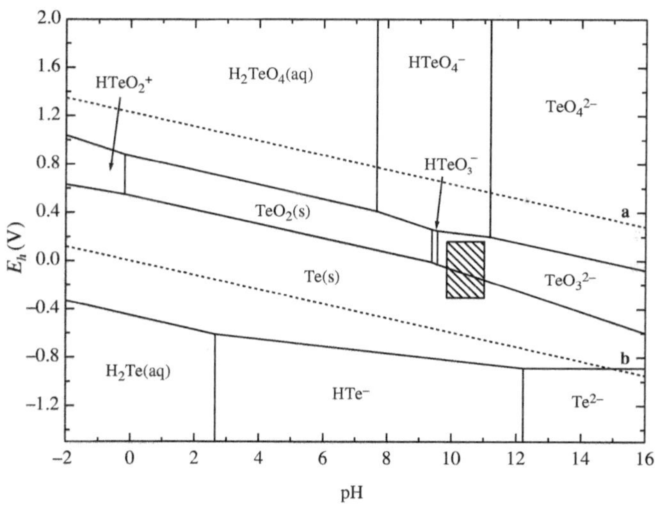

Figure 5.14 pH–potential diagram for the tellurium–water system [(Te) = 0.019 mol L^{-1}].

5.4.1.4 Zinc precipitation

Gold and silver can be recovered from the cyanide liquor by precipitation using zinc dust. The zinc precipitate will contain an extremely fine grained mass of secondary gold-bearing naumannite [(**Ag**,**Au**,Pb)$_2$(**Se**,Te), fischesserite (Ag_3AuSe_2), wulfingite ($Zn(OH)_2$)], and some zincite (ZnO).

The secondary naumannite phase that forms is similar to that observed during cyanidation, but in this step contains elementally bound gold and silver, relatively low tellurium (0.5%) and barely detectable lead, within the selenide matrix. All of the polonium present in the cyanide liquor will report to the zinc precipitate, as will most of the tellurium (94%), but only 35% of the selenium. The majority of lead will also remain with the zinc precipitate.

Cementation was used to recover gold and silver from the cyanide liquor. Reaction (5.2) describes the removal of the metals by displacement using zinc.

$$Au(CN)_2^- + Ag(CN)_2^- + Zn(s) \rightleftharpoons Au(s) + Ag(s) + Zn(CN)_4^{2-} \quad (5.2)$$

The pH–potential diagram for polonium under the operating conditions of zinc precipitation (see Fig. 5.5: box shown for zinc precipitation) at a concentration of 2×10^{-14} mol L^{-1} (Fig. 5.15) indicates that the anionic polonium species present in the cyanide solution [i.e., $H_2PoO_3(aq)$] will also reduce and form predominantly the soluble species HPo^-. In the presence of soft metals, such as lead and silver, HPo^- can react to form an insoluble polonide, for example, PbPo(s), as described by reaction (5.3).

$$HPo^- + Pb(s) + H_2O \rightleftharpoons PbPo(s) + OH^- + H_2 \quad (5.3)$$

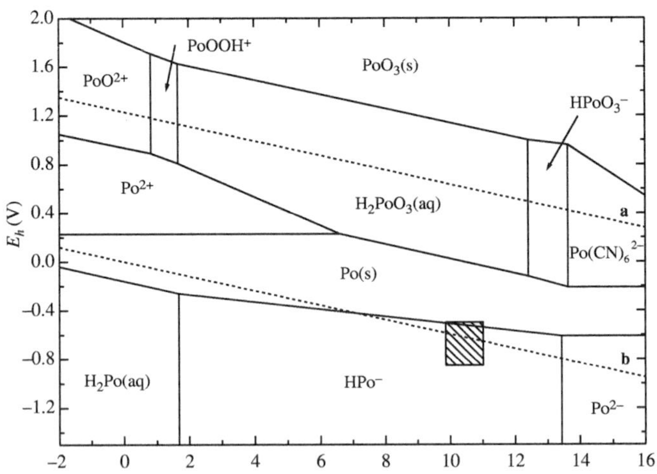

Figure 5.15 pH–potential diagram for the polonium–cyanide–water system [(Po) = 2.0×10^{-14} mol L^{-1}; (CN) = 0.71 mol L^{-1}].

which has a Gibbs energy of reaction of -160.13 kJ mol^{-1} (or equivalently, a log K of 28.1). This indicates that polonium will precipitate as an insoluble polonide in the conditions of the zinc precipitation process and the results in Table 5.3 confirm the solid phase speciation of polonium in the zinc precipitate.

The behavior of lead ions during zinc precipitation has been described by Smith and Mudder (1991); in the liquor it undergoes reduction to metallic lead. Lead was not detected in the cyanide liquor either before or after zinc precipitation. Although traces of lead can be identified during mineralogical examination of the zinc precipitate, it is difficult to identify definitive solid phases due to the small concentration of lead during zinc precipitation. Nevertheless, the pH–potential diagram for lead at a concentration of 4.5×10^{-4} mol L^{-1} under the range of operating conditions (Fig. 5.16) indicates that lead will have solid phase speciation and be present as metallic Pb(s).

Hall (1993) has stated that a large quantity of selenium is precipitated during the zinc precipitation stage, but no reference was made regarding the behavior of tellurium. Table 5.3 indicates that 35% of the selenium and 94% of the tellurium originally present in the cyanidation liquor are present in the zinc precipitate as naumannite and fischesserite (Table 5.4).

The pH–potential diagrams for the selenium– and tellurium–water systems at concentrations of 0.073 and 0.0086 mol L^{-1}, respectively, have

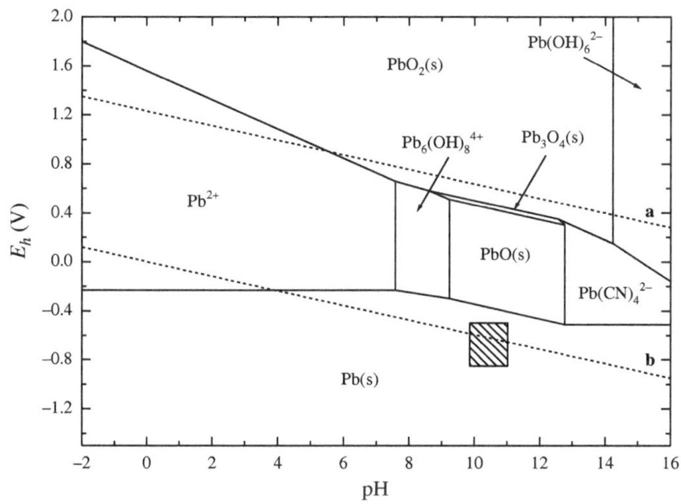

Figure 5.16 pH–potential diagram for the lead–cyanide–water system [(Pb) = 4.5×10^{-4} mol L^{-1}; (CN) = 0.71 mol L^{-1}].

again been used to predict the likely behavior of these elements under the operating conditions of zinc precipitation. Selenium (Fig. 5.17) is shown to be present predominantly as HSe^-, therefore confirming its observed behavior, occurring partially in the aqueous solution or as selenide phases such as naumannite or fischesserite. Fig. 5.18 indicates that tellurium will

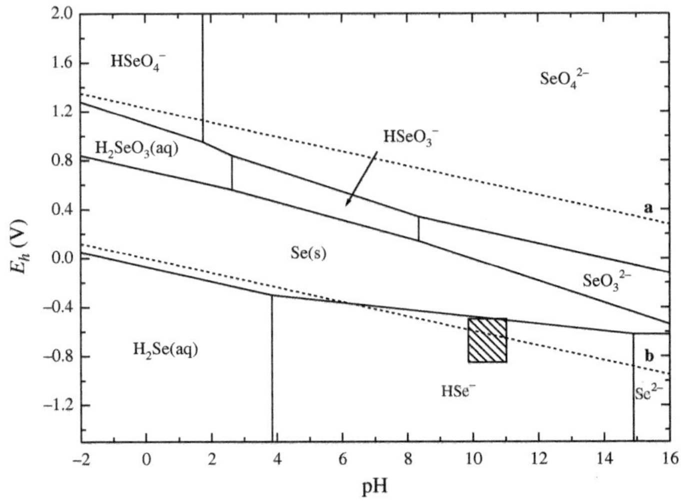

Figure 5.17 pH–potential diagram for the selenium–water system [(Se) = 0.073 mol L^{-1}].

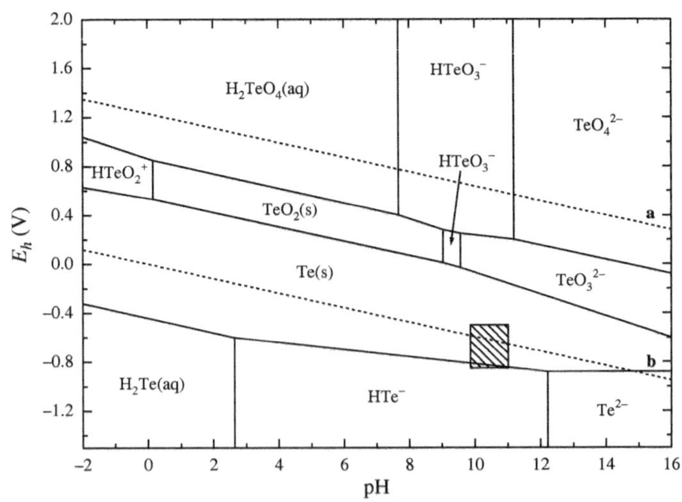

Figure 5.18 pH–potential diagram for the tellurium–water system [(Te) = 0.0086 mol L^{-1}].

be present predominantly in the elemental form, but potentially with a small amount remaining, depending on the pH–potential conditions, in the liquor as HTe^-. The presence of this latter species would likely favor the presence of tellurium as inclusions in naumannite, as is observed.

5.4.1.5 Aciding and silver recovery

The aciding process is somewhat complex, involving many chemical reactions. It is a closed system operated over three stages. In the first stage, the excess zinc metal used in zinc precipitation is removed by leaching in dilute sulfuric acid. Selenium and tellurium are then dissolved in stage two by oxidation using hydrogen peroxide and sodium nitrate. Silver dissolves in the nitric acid that forms as a result of this oxidation step and is subsequently recovered by precipitation using sodium chloride in stage three. The pH and E_h of the stage three liquor have a similar range to that of the decopperization process.

The precipitate remaining after aciding/silver recovery consists of a simple mixture of small, rounded masses of silver chloride (AgCl) agglomerates with lesser amounts of very finely dispersed metallic gold. The solid contains the majority of the polonium (94%) and lead (97%), found in the original zinc precipitate and <2% of the selenium (see Table 5.3). Analysis of the AgCl solid (using SEM/EDS) could not detect any selenium and tellurium (see Table 5.4).

In the first stage of aciding, excess zinc is dissolved in dilute sulfuric acid according to reaction (5.4).

$$Zn(s) + H_2SO_4(aq) \rightleftharpoons ZnSO_4(aq) + H_2(g) \quad (5.4)$$

Polonium would not be expected to dissolve under these conditions, analogous to its behavior during decopperization, although the pH–potential diagram at a concentration of 1.0×10^{-13} mol L^{-1} (based on the operating conditions; Fig. 5.19) indicates aqueous speciation as Po^{2+} (again, however, polonium is likely to be present within insoluble lead sulfate). Table 5.3 also indicates that polonium is present as a solid species in the AgCl and its nonleachability is again likely to be due to siting within the matrix and/or kinetics. In this case, however, zinc precipitation and then aciding can be described as an essentially continuous operation and so, there will be no time for ingrowth of ^{210}Po from its ^{210}Pb parent. If the nonleachability is due to the siting of polonium, it must be present in solid solution (e.g., lead–polonium sulfate).

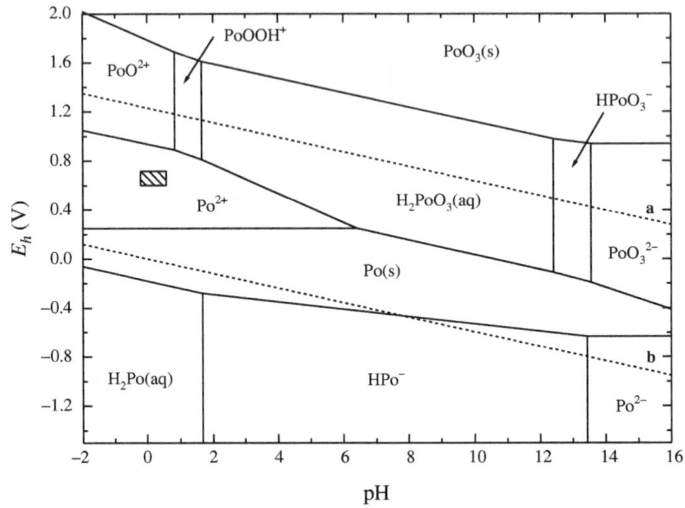

Figure 5.19 pH–potential diagram for the polonium–sulfur–water system [(Po) = 1.0×10^{-13} mol L^{-1}; (S) = 1.0×10^{-3} mol L^{-1}].

Stanley (1987) reported that lead is converted to insoluble lead sulfate during the first stage of aciding. However, lead cannot be detected by either SEM/EDS or XRD. The results in Table 5.3 and Fig. 5.20 (derived for a lead concentration of 8.2×10^{-4} mol L^{-1}) indicate that lead will be present in the AgCl solid as $PbSO_4(s)$, which supports Stanley's findings.

In the second stage of aciding, selenium and tellurium are removed from the zinc precipitate by oxidation using hydrogen peroxide and sodium nitrate. Eqs. (5.5) and (5.6) describe the overall reactions.

$$Se^{2-} + H_2O_2 + 3H^+ + O_2(g) \rightleftharpoons HSeO_4^- + 2H_2O \qquad (5.5)$$

$$Te^{2-} + 2H_2O_2 + 2H^+ + O_2(g) \rightleftharpoons H_2TeO_4(aq) + 2H_2O \qquad (5.6)$$

Since the oxidation/reduction potential for the reduction of hydrogen peroxide (reaction 5.7) is 1.776 V (Vanýek, 1996), it has been assumed that, during this stage, there is a significant increase in the E_h.

$$H_2O_2 + 2H^+ + 2e^- \rightleftharpoons 2H_2O \qquad (5.7)$$

The pH–potential diagrams for the selenium– and tellurium–water systems, have been derived for concentrations of 0.13 and 0.016 mol L^{-1}, respectively. The pH range used for polonium and lead during aciding is

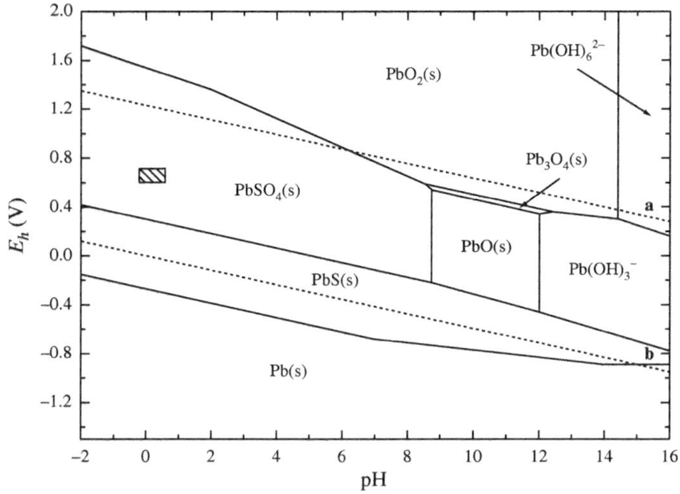

Figure 5.20 pH–potential diagram for the lead–sulfur–water system [(Pb) = 1.0×10^{-3} mol L^{-1}; (S) = 1.0×10^{-3} mol L^{-1}].

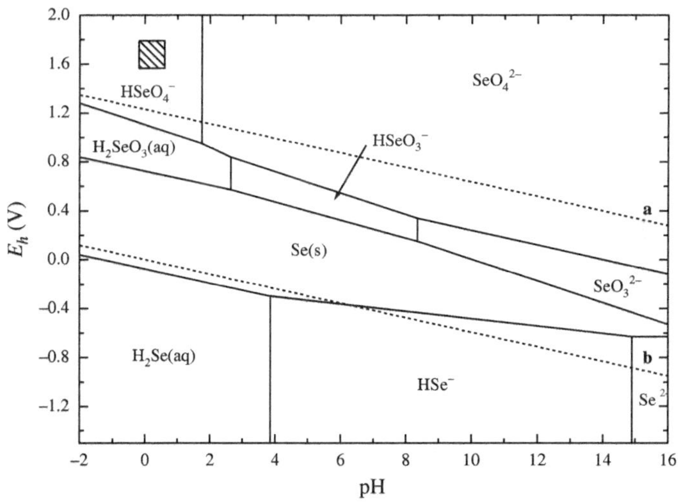

Figure 5.21 pH–potential diagram for the selenium–water system [(Se) = 0.13 mol L^{-1}].

retained also for selenium and tellurium. It is, acknowledged that the E_h, although elevated, is most likely not as high as 1.776 V and a range from this value down to 1.576 V (therefore the range in E_h is the same as utilized for polonium and lead in this stage) has been used (Figs. 5.21

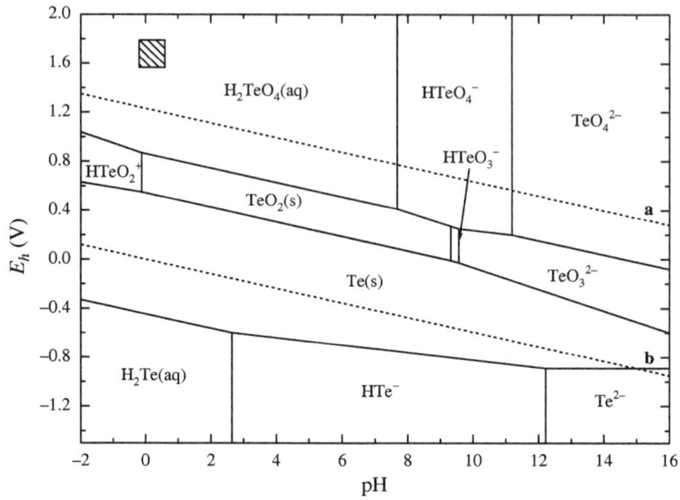

Figure 5.22 pH−potential diagram for the tellurium−water system [(Te) = 0.016 mol L^{-1}].

and 5.22). The pH and potential ranges confirm the aqueous speciation for selenium and tellurium as given in Eqs. (5.5) and (5.6), respectively.

In stage three of the process, silver is dissolved by the nitric acid that forms in the oxidation and is subsequently recovered as insoluble silver chloride by the addition of sodium chloride to the dilute nitric acid solution. Eqs. (5.8) and (5.9) describe the reactions.

$$2Ag(s) + 2HNO_3 \rightleftharpoons 2AgNO_3 + H_2(g) \tag{5.8}$$

$$AgNO_3 + NaCl(s) \rightleftharpoons AgCl(s) + NaNO_3 \tag{5.9}$$

Table 5.3 indicates that the AgCl solid contains selenium; however, the element was not detected using SEM/EDS. The likely speciation of selenium in this solid is Se(s), as shown in Fig. 5.23 (pH and E_h ranges similar to those of the decopperization process). Although Table 5.3 indicates that there is no tellurium present in the AgCl solid, Hall (1993) suggests that trace tellurium may be present. Like selenium, this was not detected by SEM/EDS analysis. Fig. 5.24 (pH and E_h ranges similar to those of the decopperization process) indicates $TeO_2(s)$ as the likely species.

5.4.2 Overview

The literature review (Chapter 2: Physical and chemical properties) indicated that the chemical behavior of polonium closely followed that of

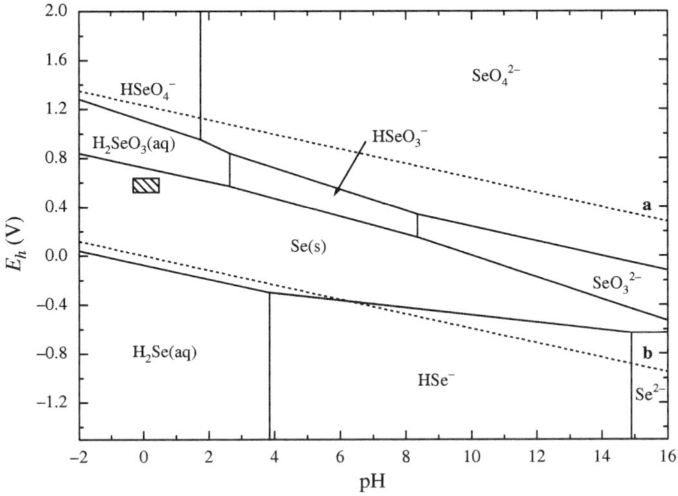

Figure 5.23 pH–potential diagram for the selenium–water system [(Se) = 0.13 mol L^{-1}].

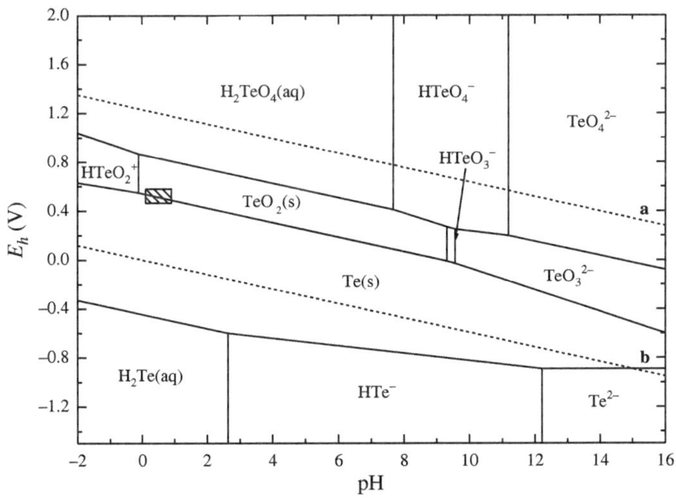

Figure 5.24 pH–potential diagram for the tellurium–water system [(Te) = 0.016 mol L^{-1}].

tellurium and, to a lesser extent, selenium, while the physical properties resembled its neighbors, lead and bismuth (Bagnall, 1957a). This work has demonstrated that the chemistry and deportment of ^{210}Po during precious metals recovery appears to be analogous to that of lead, in particular, during the zinc precipitation and aciding/silver recovery stages. The major findings are summarized as follows:

1. The phases containing Po, Pb, Se, and Te in the raw slimes do not leach during decopperization using sulfuric acid, although aqueous speciation is indicated on the pH–potential diagram for polonium.
2. The phases containing Se and Te dissolve in cyanidation whereas the phases containing Po and Pb do not.
3. Polonium, Pb, and Se form solid phases during zinc precipitation whereas Te remains in solution.
4. Polonium and Pb remain as solid phases during aciding/silver recovery. The solid phases containing Se and Te dissolve during the oxidation stage.
5. The nonleachability of polonium is most likely due to: (a) siting within matrices not amenable to leaching; (b) dissolution and then coprecipitation within a solid phase not amenable to leaching; and/or (c) kinetics.

5.4.3 Oxidation experiments

5.4.3.1 Oxidation using calcium hypochlorite

To gain a better understanding of the behavior of polonium in relation to lead during the decopperization and cyanidation stages, additional studies were carried out by Brown (2001).

The aim of these experiments was to study the effect of oxidation on the dissolution behavior of polonium and lead at a constant pH and varying E_h. In an initial experiment, recently produced decopperized anode slimes were cyanided over 6 hours by first neutralizing a 10% *w*/*v* slurry pH = 2.16–7.5 by the addition of sodium hydroxide, and then adding sodium cyanide solution (20%) to achieve, and thence maintain, a free cyanide concentration of 1%. After steady-state conditions were attained, the average pH and E_h values for the cyanidation were 11.84 and 0.081 V, respectively.

The free cyanide was determined at regular intervals throughout the process by titration with silver nitrate according to the method of Vogel (1961a). At the conclusion of cyanidation, a sample was taken and filtered through a 0.45 μm filter. The liquor was retained while the solid was washed thoroughly with deionized water, dried at 100°C and weighed.

Calcium hypochlorite solution (10% available chlorine) was then added to the cyanided slurry (pH = 11.97; E_h = 0.044 V) via a burette until an E_h of 0.160 V was reached. After this potential was maintained for 15 minutes (pH = 10.26), a sample (ClO_2 1) was taken and filtered through a 0.45 μm filter. The liquor was retained for analysis. Calcium hypochlorite was again added until an E_h of 0.260 V was reached. After this potential was maintained for a further 15 minutes (pH = 9.54), a sample (ClO_2 2) was taken and processed as described previously. Further calcium hypochlorite was added to attain a target E_h of 0.360 V, but the pH dropped dramatically. Because of the danger of generating hydrogen cyanide, the experiment was stopped and sodium hydroxide added to increase the pH. The final slurry pH and E_h values were 8.02 and 0.054 V, respectively. A sample (ClO_2 Final) of the final slurry was processed in the same manner as the cyanided slurry.

The liquors were analyzed using ICPAES and the solids analyzed (as fused glass disks) using a Philips PW2400 wave dispersive X-ray fluorescence spectrometer (XRF). Polonium-210 in the solid and liquor samples was determined by the method described in Section 5.7. Lead-210 in these samples was separated by solvent extraction using 1% diethyl ammonium diethyl dithiocarbamate (DDTC) in chloroform from 1.5 mol L^{-1} hydrochloric acid, made by diluting fourfold the aqueous phase remaining after ^{210}Po extraction. The extract was evaporated and then ^{210}Pb was precipitated as the chromate from a buffered acetate solution at pH 4.5. The lead chromate was counted using a Canberra 2404 Alpha–Beta–Gamma-Spectrometer and the activity of ^{210}Pb calculated from the measured count rate of its daughter, ^{210}Bi. Table 5.5 contains a summary of the results.

Table 5.5 Results summary for the oxidation of cyanided anode slimes using calcium hypochlorite.

	Liquors (mg L^{-1})[a]				Solids (%)[b]		
	Cyanide solution	**ClO_2 1**	**ClO_2 2**	**ClO_2 final**	**Decopp. slimes**	**Cyanided solid**	**ClO_2 solid**
pH	**11.97**	**10.26**	**9.54**				
E_h	**0.044 V**	**0.160 V**	**0.260 V**				
Se	11,000	15,000	10,000	5700	11	0.93	1.4
Te	220	120	0.0014	<0.7	0.70	0.83	1.1
Bi	<0.8	<0.8	<0.8	<0.8	0.41	0.59	0.57
Pb	9.7	0.90	0.50	<0.4	7.0	10	9.9
^{210}Po	14.4	1100	1000	324	1440	2020	1540
^{210}Pb					2280	2930	2200

[a] ^{210}Po is Bq L^{-1}.
[b] ^{210}Po and ^{210}Pb are Bq g^{-1}.

5.4.3.2 Oxidation using sodium hypochlorite

The dramatic fall in the pH had not been foreseen and so a further experiment was carried out using the same procedure as outlined above except for the following:

1. Sodium hypochlorite (13% available chlorine) was used as the oxidant.
2. Sodium hydroxide (5%) was used to control the pH at ~11.
3. The addition of reagents was controlled using a Metrohm titration unit.

The pH and E_h of the cyanidation liquor were 11.43 and 0.062 V, respectively. Sample 1 (OCl 1) was taken at a potential of 0.140 V (pH = 11.09), sample 2 (OCl 2) at 0.210 V (pH = 11.09) and sample 3 (OCl 3) at approximately 0.245 V (pH ~ 11.0). Above this potential, it was not possible to maintain a constant pH. As hypochlorite was added, the potential increased and the pH decreased. If more hydroxide was added to increase the pH, the potential decreased. The experiment was then stopped.

Analyses were carried out on solid and liquor samples as outlined previously and Table 5.6 contains a summary of the results.

Mass balance data for the cyanidation stages of each experiment and that from the cyanidation section in anode slimes processing (see above) are summarized in Table 5.7. Dissolutions during cyanidation were comparable to those obtained in the plant with the exception of tellurium and

Table 5.6 Results summary for the oxidation of cyanided anode slimes using sodium hypochlorite at constant pH.

	Liquors ($mg\ L^{-1}$)[a]				Solids (%)[b]		
	Cyanide solution	OCl 1	OCl 2	OCl 3	Decopp. slimes	Cyanided solid	OCl solid
pH	11.43	11.09	11.09	11.00			
E_h	0.062 V	0.140 V	0.210 V	0.245 V			
	13,000	12,000	12,000	8500	11	0.76	0.76
Te	190	77	35	8.0	0.70	0.90	1.1
Bi	<0.8	<0.8	<0.8	<0.8	0.55	0.60	0.59
Pb	2.8	<0.40	0.40	2.5	7.1	11	11
^{210}Po	10.8	151	160	7.6	1250	2260	2600
^{210}Pb					2280	4500	4730

[a] ^{210}Po is $Bq\ L^{-1}$.
[b] ^{210}Po and ^{210}Pb are $Bq\ g^{-1}$.

the reason for this is not known. Possible reasons for the difference observed during these experiments are: (1) reduced reaction time (6 vs 8 hours in the plant) and (2) mechanical stirring rather than air-sparging (plant).

Mass balance data for the oxidation of cyanided anode slimes using sodium hypochlorite are summarized in Table 5.8. Although there were difficulties in trying to maintain a constant pH for increasing redox potential, the data clearly show a similarity between the behavior of polonium and lead compared with selenium and, to a much lesser extent under these conditions, tellurium.

5.4.4 Acid leaching experiments

The aim of these experiments was to study polonium and lead chemistry under controlled conditions. The mass balance data generated in anode slimes processing (Section 5.4.1) were based on production figures, and hence, dealt with bulk quantities of materials, for example, tonnes and kiloliters. To study similarities between the chemistry of polonium and lead in anode slimes processing more accurately, a series of acid leaching experiments were carried out on recently produced decopperized anode slimes and on a sample of aged decopperized anode slimes generated from a study 3 years prior. The term "aged" is used here to describe the condition where unsupported ^{210}Po has decayed and ^{210}Po activity is due solely to secular equilibrium existing between ^{210}Po and its parent, ^{210}Pb. In recently produced material, secular equilibrium will not necessarily exist and unsupported ^{210}Po may be present in different phases within the matrix. Under these conditions, differences in leaching behavior may be observed because of siting characteristics. The respective average concentrations of ^{210}Po and ^{210}Pb in each of the solids were:

recently produced decopperized slimes—2351 Bq g^{-1} (Po); 3004 Bq g^{-1} (Pb) and

aged decopperized slimes—3391 Bq g^{-1} (Po); 3544 Bq g^{-1} (Pb).

5.4.4.1 *Sulfuric acid leaching*

Acid leaching was carried out in triplicate on the recently produced material and in duplicate on the aged slimes because of a limited quantity at hand. In all experiments, the solids were leached for 23 hours, uncovered, using 0.5 mol L^{-1} sulfuric acid at a slurry density of 10% *w*/*v*. The solutions were stirred constantly and the temperature maintained at 25°C by the use of thermostatically controlled water baths. The average pH of the

Table 5.7 Mass balance data for cyanidation of anode slimes (% dissolution based on liquor analysis).

Element	Exp. 1	Exp. 2	
Se	100	100	100
Te	36	31	100
Bi	<0.2	<0.2	<0.5
Pb	0.05	0.05	<0.5
^{210}Po	0.01	0.01	0.6

Table 5.8 Mass balance data for oxidation of cyanided anode slimes using sodium hypochlorite (% dissolution based on liquor analysis).

Element	OCl 3 liquor
Se	103
Te	1.6
Bi	<0.2
Pb	0.05
^{210}Po	0.01

initial slurries was 0.19 (range 0.18–0.20) and that of the final slurries, 0.29 (range 0.27–0.33). The average E_h during leaching was 0.748 V (range 0.735–0.761 V).

After leaching, the slurries were filtered through 541 Whatman filter papers and the liquors retained. The solids were displacement washed using an equivalent volume of deionized water, dried at 100°C and weighed. The solids were then pulverized. The liquors were filtered through a 0.45 μm filter paper.

The liquors were submitted for analysis using ICPAES and the solids analyzed using XRF. Polonium-210 and ^{210}Pb in the solid and liquor samples were determined using α- and β-spectrometry, respectively (see Section 5.4.3). Table 5.9 contains a summary of the results.

The dissolution of all elements, except polonium, was found to be consistent between the two materials. Although polonium dissolution was a factor of 30 times greater in recently produced decopperized slimes compared to the aged material, >99.5% of the polonium remained in the solid. The most likely explanation for this variation is surface adsorption (see below).

Table 5.9 Results summary for sulfuric acid leaching of recently produced and aged decopperized anode slimes (pH = 0.29; E_h = 0.748 V) (% dissolution based on liquor analysis).

	Decopperized anode slimes				
	Recent			Aged	
	A	B	C	E	F
Se	< 0.005	< 0.005	< 0.005	< 0.005	< 0.005
Te	3.0	3.2	3.2	4.5	4.8
Bi	8.7	9.0	9.0	13	14
Pb	0.050	0.050	0.051	0.047	0.048
^{210}Po	0.32	0.27	0.28	0.012	0.015
^{210}Pb	0.054	0.059	0.058	0.050	0.055

5.4.4.2 Leaching with different leachants

A series of experiments were then carried out to study the effects of different leachants on dissolution. Four 20 g samples of recently produced decopperized anode slimes were leached using the following leachants under the same conditions as those used for the 0.5 mol L^{-1} sulfuric acid leach:

1. 0.5 mol L^{-1} H_2SO_4;
2. 1 mol L^{-1} H_2SO_4;
3. 1 mol L^{-1} HCl; and
4. 1 mol L^{-1} HCl/1 mol L^{-1} $MgSO_4$.

After leaching, samples were separated and analyzed as detailed above and Table 5.10 summarizes the results.

Polonium and lead dissolution increased slightly with an increase in sulfuric acid concentration, however, >99.5% of each element again remained in the solid. The dissolution of both elements increased significantly in the hydrochloric acid leach (~100-fold), however, this was reduced to ~30- and ~10-fold, respectively, for the HCl/SO_4 mixture. The pH–potential diagrams for the polonium–chlorine– and lead–chlorine–water systems (Figs. 5.25 and 5.26) indicate that the aqueous species present are $PoCl_4^{2-}$ and $PbCl_2(aq)$, respectively.

Table 5.4 indicates that the phase in the decopperized slimes containing the majority of the lead is anglesite ($PbSO_4(s)$), with eucairite containing lesser amounts. Since $PbSO_4(s)$ is only slightly more soluble in hydrochloric acid than in sulfuric acid, the observed lead dissolution is most likely due to leaching from the eucairite matrix. In comparison to

Table 5.10 Results summary for leaching of recently produced decopperized anode slimes using various leachants (% dissolution based on liquor analysis).

	0.5 mol L^{-1} H_2SO_4	1 mol L^{-1} H_2SO_4	1 mol L^{-1} HCl	1 mol L^{-1} HCl/1 mol L^{-1} $MgSO_4$
pH	0.29	0.03	0.06	0.06
E_h	0.608 V	0.738 V	0.637 V	0.629 V
Se	< 0.005	< 0.005	< 0.005	< 0.005
Te	2.5	2.8	2.5	2.4
Bi	7.2	13	46	29
Pb	0.052	0.056	11	0.62
^{210}Po	0.33	0.45	50	13
^{210}Pb	0.059	0.065	9.6	0.64

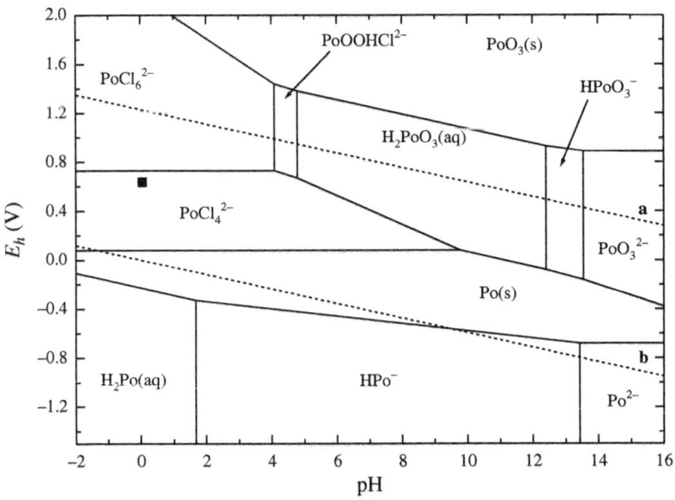

Figure 5.25 pH–potential diagram for the polonium–chlorine–water system [(Po) = 5.6×10^{-12} mol L^{-1}; (Cl) = 1 mol L^{-1}].

lead, the greater solubility of polonium is due to the increased stability of its chloride complexes (see Chapter 3: Chemical thermodynamics of polonium). Dissolution of both elements is suppressed in the HCl/SO_4 mixture due to the increased sulfate concentration, and therefore the ability of either lead or polonium to be released from the anglesite matrix is reduced.

Conversely, selenium and tellurium do not dissolve during either sulfuric or hydrochloric acid leaching. The pH–potential diagrams for the

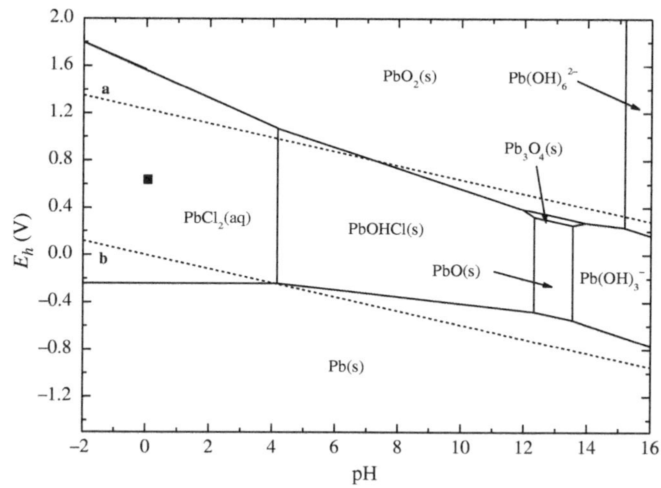

Figure 5.26 pH–potential diagram for the lead–chlorine–water system [(Pb) = 0.034 mol L^{-1}; (Cl) = 1 mol L^{-1}].

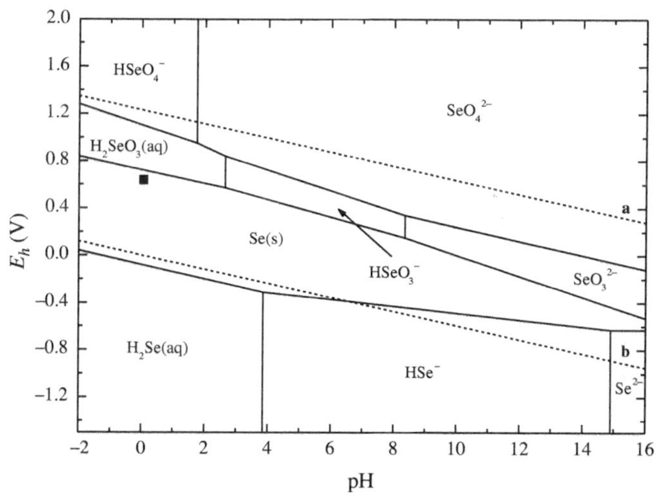

Figure 5.27 pH–potential diagram for the selenium–water system [(Se) = 0.15 mol L^{-1}; (Cl) = 1 mol L^{-1}].

selenium– and tellurium–water systems in sulfate media (as shown previously in Figs. 5.23 and 5.24) and chloride media (Figs. 5.27 and 5.28) confirm these results and indicate solid phase speciation as Se and TeO_2 in both the sulfate and chloride systems.

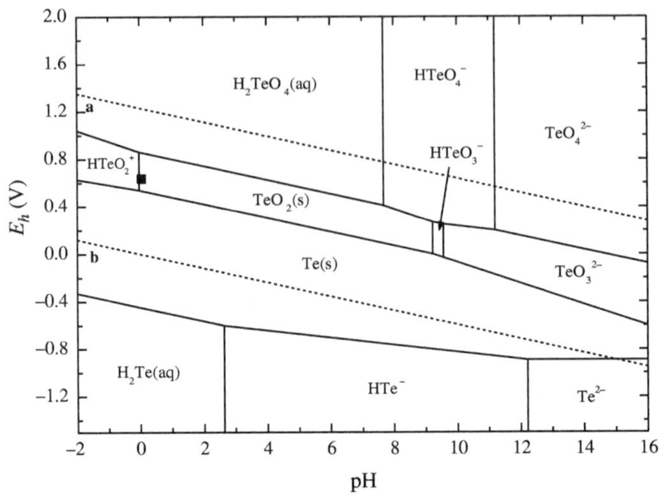

Figure 5.28 pH–potential diagram for the tellurium–water system [(Te) = 0.013 mol L^{-1}; (Cl) = 1 mol L^{-1}].

5.4.5 Preparation of lead/polonium sulfate

The results obtained in previous sections have indicated that the siting of polonium within the anode slimes material is most likely with the lead phases, primarily lead sulfate, which has been identified by XRD/SEM analysis and has been shown not to leach in sulfuric acid processes.

To study polonium behavior in relation to lead in a sulfate system, lead sulfate was prepared according to the method of Vogel (1961b). Lead nitrate was accurately weighed and dissolved in deionized water. A known amount of ^{209}Po tracer was then added. Concentrated sulfuric acid (2 mL) was added to the solution, with stirring, and the mixture evaporated until thick white fumes of sulfuric acid were freely evolved. After cooling, the mixture was diluted to 40 mL and left standing for 1 hour. The final solution pH and E_h were 0.23 and 0.692 V, respectively. The precipitate was filtered through a weighed filter apparatus, washed several times with 1% sulfuric acid and dried in an oven at 75°C until a constant weight was achieved. The percentage of lead sulfate produced was then calculated. The supernatant and all washings were retained.

A polonium analysis was carried out on the retained solution in the same manner as described previously. From the measured count rate, the amount of activity on the disk, and hence the amount of activity reporting to the lead sulfate precipitate, were calculated. The results are summarized in Table 5.11.

Table 5.11 Polonium behavior in the preparation of lead sulfate [(^{209}Po) = 9.1 × 10^{-14} mol L^{-1}].

% $PbSO_4$ produced	% ^{209}Po on counting disk	% ^{209}Po in $PbSO_4$
99.7	25.4	74.6

The results show that a significant proportion of the added polonium coprecipitates with the lead sulfate, which supports the view that polonium is associated with this phase in the anode slimes.

Coprecipitation describes the process by which a normally soluble component of a solution is carried down during the formation of a precipitate (Skoog and West, 1982). Two important types of coprecipitation are surface adsorption and occlusion (Vogel, 1961c). Surface adsorption describes adsorption at the surface of the particles exposed to the solution. For precipitates with ionic lattices, the ion that is most strongly adsorbed by a crystal lattice is that ion which forms the least soluble salt. The deformability of the adsorbed ions and the electrolytic dissociation of the adsorbed compound also have a significant influence.

Occlusion can be described as occurring during the build-up of a precipitate from the primary particles. There will always be a certain amount of surface adsorption and, during coalescence, the impurities will be partially eliminated if large single crystals are formed and the process takes place slowly. Alternatively, if coalescence is rapid, large aggregates composed of loosely bound small crystals with impurities being entrained at interfaces will form. If the impurity is isomorphous or forms a solid solution with the precipitate, the amount of coprecipitation may be very large since there will be no tendency for elimination during the aging process.

Solid solutions are solids in which one component is randomly dispersed, at an atomic or molecular scale, throughout another component. As in any crystal, the packing of atoms is ordered, but there is not necessarily any particular order as to which lattice points are occupied by which type of atom (Sienko and Plane, 1976). Minerals can be described as solid solutions (Garrels and Christ, 1965). When a solid solution is formed by substituting one cation for another, the extent to which solution takes place and the stability of the resulting solution depend upon the likeness of the two cations. If the substituent cation has the same charge, coordination geometry preference and nearly the same ionic radius as the dispossessed cation, then a stable solid solution takes place readily.

It is difficult to ascertain which of the above processes best describes the formation of the lead/polonium sulfate. Potentially, both coprecipitation and occlusion describe the observed behavior. For solid solution to occur, however, the ionic radius of the Po^{2+} ion would need to be around 0.98 Å (Shannon, 1976), the ionic radius of Pb^{2+}. Unfortunately, no value for the former parameter was found in the literature and so formation of the sulfate via solid solution is not conclusive.

5.4.6 Discussion

The results from the above studies clearly demonstrate that the behavior of polonium during anode slimes processing is similar to lead, highlighted by the chemical similarities in many of the stages, as compared with selenium and tellurium. Apart from indicating that coprecipitation of polonium with lead is feasible, the results give no other indications of the possible siting of polonium within the phases. It seems likely that polonium enters the anode slimes with the lead sulfate produced as a result of copper electrorefining, which uses copper sulfate/sulfuric acid as the electrolyte.

The pH−potential diagrams indicated aqueous speciation for polonium in sulfate media, however, that was not observed in practice. The three possible reasons proposed for this behavior were: (1) siting within matrices not amenable to leaching; (2) dissolution and then coprecipitation within a solid phase not amenable to leaching; and/or (3) kinetics. Since it is probable that polonium partitions with lead sulfate during anode slimes processing, the conclusion drawn is that polonium is not amenable to leaching because of siting within the matrix and hence remains with the lead.

Of interest are the sulfuric acid leaching results for the aged and recently produced materials. Lead dissolution was constant between the two materials; however, polonium dissolution, although still very minimal, was a factor of 30 times greater in the "recently produced" material. This is most likely due to activity at the surface being more accessible to leaching.

The rates of chemical reactions and the mechanisms by which they occur are described by chemical kinetics. It is not possible to predict the reaction kinetics of polonium during any stage of anode slimes processing, based on the thermodynamic data and experimental procedures used in this section. Clearly, extensive work beyond the scope of that presented would have to be undertaken to address this issue.

5.5 Case study—polonium behavior during silver and gold electrorefining

The acided zinc precipitate produced during anode slimes processing is smelted to produce a doré button containing 80%–90% silver and 5%–15% gold. This button is then remelted, oxidized to remove lead and cast into anodes (Hall, 1993). Silver is recovered from the doré anode by electrorefining using silver nitrate as the electrolyte; silver is dissolved and then plated out as silver crystal onto the cathode. Polonium-210 enters the circuit during the dissolution process and must be removed to ensure that product specifications are met. This is achieved by circulating the electrolyte through an activated carbon filter to adsorb polonium, although the mechanism for removal is not known.

In the polonium–nitrate system, the complexes that form are weak and under the conditions prevailing in silver electrorefining, polonium–nitrate complexes are not dominant. As such, polonium behavior is described by the pH–potential diagram for the polonium–water system (Fig. 5.29), derived for a polonium concentration of 6.5×10^{-14} mol L^{-1}. The diagram indicates that polonium forms aqueous H_2PoO_3 under the operating conditions (pH = 4.16; E_h = 0.596 V).

Activated carbon is a microcrystalline, nongraphitic form of carbon characterized by a large specific surface area, typically 300–2500 m^2 g^{-1}, which allows the physical adsorption of gases and vapors from gases and

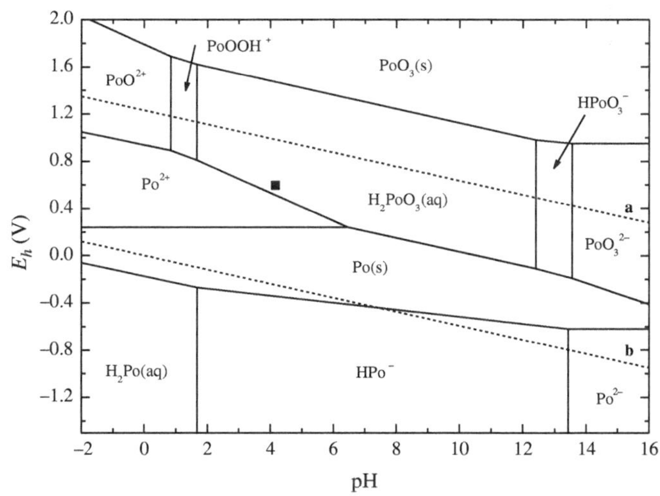

Figure 5.29 pH–potential diagram for the polonium–water system [(Po) = 6.5×10^{-14} mol L^{-1} in 0.024 mol L^{-1} nitrate].

dissolved or dispersed substances from liquids. Liquid-phase adsorbents have a majority of pores ≥3 nm; larger pores being required because of the need for rapid diffusion in the liquid and also because of the large size of many dissolved adsorbates. One of the most important chemical properties of activated carbon is the pH, which measures the surface acidity or basicity of the oxygen-containing groups. This assists in predicting hydrophilicity and anionic or cationic adsorptive preferences (Grayson, 1978).

The adsorption of metal ions onto activated carbon varies markedly with pH and the onset of extensive adsorption coincides, in most cases, with the first stage of hydrolysis of the hydrated metal ion (Kadirvelu et al., 2000). Since polonium is present as a hydrolyzed species, PoO $(OH)_2$ (i.e., H_2PoO_3), in electrorefining, it is likely to be adsorbed by the carbon in a similar manner. Conversely, lead in the electrolyte is not adsorbed by the activated carbon column (Brown, 2001). In a similar manner to that described for polonium above, lead–nitrate complexes do not dominate under the conditions of silver electrorefining. The pH–potential diagram for the lead–water system (Fig. 5.30) indicates that lead is not hydrolyzed and is present as Pb^{2+}. Kadirvelu et al. (2000) also observed minimal adsorption of the Pb^{2+} ion by activated carbon.

The slimes (gold mud) from silver electrorefining contain the gold. The gold mud is leached with nitric acid, to remove any remaining silver, and then sulfuric acid, to remove trace selenium and tellurium, to give a

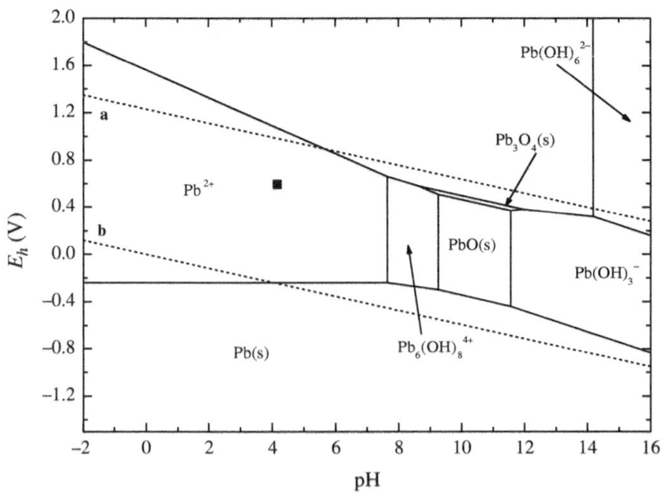

Figure 5.30 pH–potential diagram for the lead–water system [(Pb) = 3.6×10^{-4} mol L^{-1} in 0.024 mol L^{-1} nitrate].

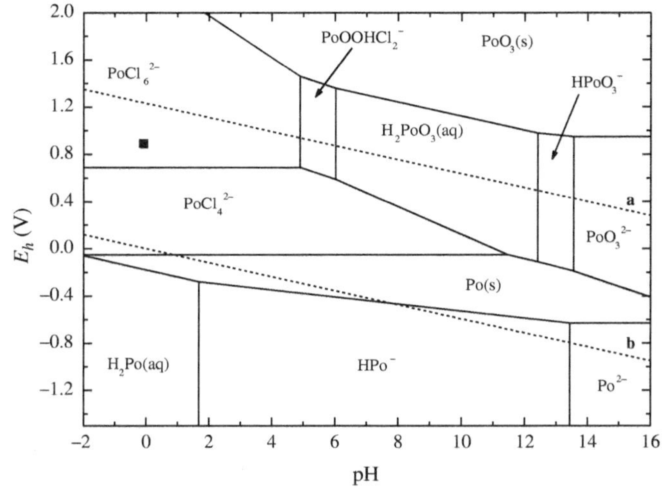

Figure 5.31 pH–potential diagram for the polonium–chlorine–water system [(Po) = 8.5×10^{-14} mol L^{-1}; (Cl) = 4.1 mol L^{-1}].

gold sponge containing 99% Au. After melting and casting into anodes, gold is recovered via a two-stage electrorefining process using gold chloride as the electrolyte (Hall, 1993).

The pH–potential diagram for the polonium–chlorine–water system (Fig. 5.31), derived for a polonium concentration of 8.5×10^{-14} mol L^{-1} in stage one electrorefining, indicates that polonium forms $PoCl_6{}^{2-}$ under the operating conditions (pH = −0.08; E_h = 0.891 V). Although polonium accumulates in the electrolytes through successive uses, it remains in solution and is not found in the cathodes.

5.6 Case study—polonium in seawater

Carbon dioxide in the atmosphere dissolves in the surface layer of the ocean and the transfer of carbon to the deep ocean is facilitated by fixation into particulate organic matter in the euphotic zone. [The top 50–120 m of the ocean layer where light penetrates contains phytoplankton and zooplankton (G. Peck, private communication).] Sinking of these biogenic particles from the upper to the deeper layers of the ocean is an important pathway in the global carbon cycle. To develop accurate carbon cycle models, particle fluxes must be quantified (Peck and Smith, 2000). Geochemical tracers allow estimation of the rates of marine processes that are difficult to measure directly. Polonium-210 and ^{210}Pb are

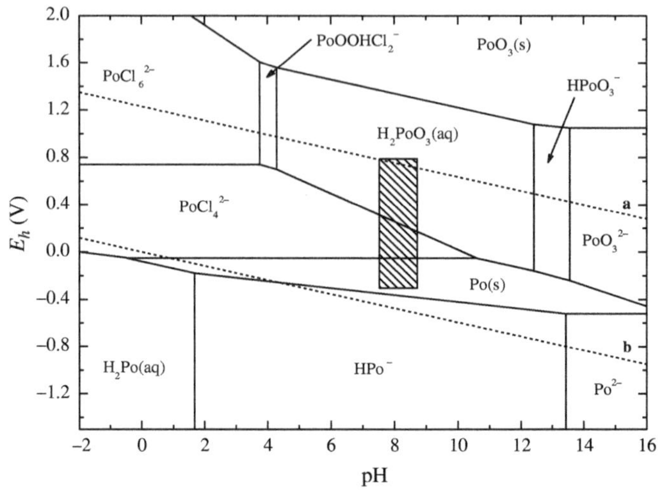

Figure 5.32 pH−potential diagram for the polonium−chlorine−water system [(Po) = 3.17×10^{-17} mol L^{-1}; (Cl) = 0.559 mol L^{-1}].

useful tracers of particle fluxes in the upper layer of the ocean (Towler and Smith, 1997) because of their relatively short half-lives [138.378 days and 22.3 years, respectively (Kocher, 1977)].

Lead-210 is added to the upper ocean by fallout from the atmosphere and by decay of ^{226}Ra in the water column. Polonium-210 is produced in seawater from the decay of ^{210}Pb. Polonium-210 and ^{210}Pb are removed from the upper ocean by both radioactive decay and adsorption onto particles followed by sinking. The input of ^{210}Pb from the atmosphere and removal by sinking particles combine to give a vertical concentration profile that can be used to calculate the ratios of the contributing processes. Measurements of the fractions of ^{210}Po and ^{210}Pb in the dissolved (<0.45 μm) and particulate (>0.45 μm) phases are used to calculate radionuclide residence times (Peck and Smith, 2000).

An understanding of polonium and lead speciation in seawater can be provided by pH−potential diagrams. This, in turn, aids the marine chemist in interpreting the distribution of these elements between dissolved, particulate, and biogenic (e.g., phytoplankton and zooplankton) phases in the ocean layers.

The pH−potential diagram for the polonium−chlorine−water system is illustrated in Fig. 5.32. It has been derived for a polonium concentration of 3.17×10^{-17} mol L^{-1} (Peck and Smith, 2000) and a chlorine

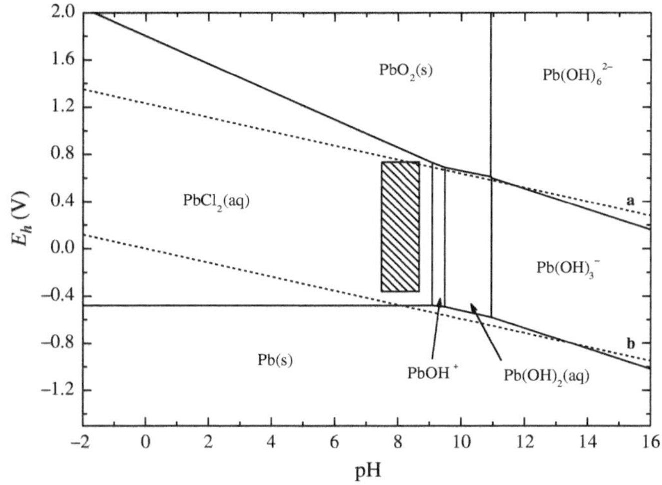

Figure 5.33 pH–potential diagram for the lead–chlorine–water system [(Pb) = 10^{-10} mol L^{-1}; (Cl) = 0.559 mol L^{-1}].

concentration of 0.559 mol L^{-1} (Ahrland, 1975). For the pH and E_h ranges described by Kester et al. (1975) [i.e., 7.3 to 8.4 and −0.3 to 0.8 V, respectively (see hatched box in the figure)], polonium exists predominantly as the aqueous species $PoCl_4^{2-}$ and H_2PoO_3(aq).

Similarly, the pH–potential diagram for the lead–chlorine–water system (Fig. 5.33), derived for a lead concentration of 10^{-10} mol L^{-1} (Ahrland, 1975), shows that lead exists predominantly as the aqueous species, $PbCl_2$(aq), which is consistent with the inorganic speciation data for seawater described by Kester et al. (1975).

In their study, Peck and Smith (2000) found that, in near-surface waters, the residence time for dissolved ^{210}Po was short (0.27 years), which suggests that polonium is rapidly adsorbed onto particles. Total ^{210}Pb is dominated by the dissolved fraction [concentrations up to 200 times greater than particulate ^{210}Pb (as is indicated in Fig. 5.33)]. The residence time for dissolved ^{210}Pb, however, was much longer than that of dissolved ^{210}Po (8.0 years).

In his review, Lyklema (1975) discusses some of the fundamental properties of charged interfaces in relation to the distribution of charge and potential under conditions prevailing in seawater. At the solid–liquid interface between seawater and dispersed particulate matter, adsorption occurs. Interfaces as a whole are always electroneutral, although an unequal distribution of ions can result in charged interfaces. The most common mechanisms by which interfaces can acquire a charge are:

1. Dissociation of surface groups resulting in either positively or negatively charged surfaces depending on the point of zero charge of the substance.
2. Preferential adsorption of ions occurring, to a large extent, only if the adsorbing species has a high affinity for the substance.
3. Isomorphic substitution, for example, Si^{4+} by Al^{3+}, typically seen in clay minerals.

It seems plausible, therefore, that the rapid removal of ^{210}Po compared to ^{210}Pb from near-surface water could be due to removal as the charged species, $PoCl_4{}^{2-}$, via mechanism (1). Alternatively, removal as the hydrolyzed species, $PoO(OH)_2(aq)$, in a similar manner to that described previously for polonium adsorption onto activated carbon, may occur via mechanism (2).

5.7 Case study—autodeposition of polonium

The analysis of polonium from within complex solids or aqueous solutions using radiochemical techniques commonly involves its autodeposition onto metals such as silver following multiple steps to separate the polonium (Brown, 2001). Ferric iron is used as a carrier for the polonium and ammonia is used to precipitate iron hydroxide which carries the polonium. The ammonia can also be used to remove copper impurities through the formation of copper tetraammine. The precipitated iron hydroxide is then dissolved using hydrochloric acid and the iron and polonium separated using sequential solvent extraction steps. After evaporation of the final organic extract, containing polonium, to dryness, hydrochloric acid is added, together with citrate to complex any remaining iron and hydroxylamine hydrochloride to reduce the E_h and precipitate any gold or selenium present in the solution (these latter elements are separated via filtration). The polonium is then autodeposited onto a silver disk under acidic (pH of 1.5) and reducing (E_h of 0.18 V) conditions. For illustration a polonium concentration of 10^{-12} mol L^{-1} has been used. In the absence of silver metal and under these pH−potential conditions, polonium remains in solution as the $PoCl_4{}^{2-}$ species, as shown in Fig. 5.34. Consequently, if no further reaction takes place, the deposition of polonium would be unlikely to occur.

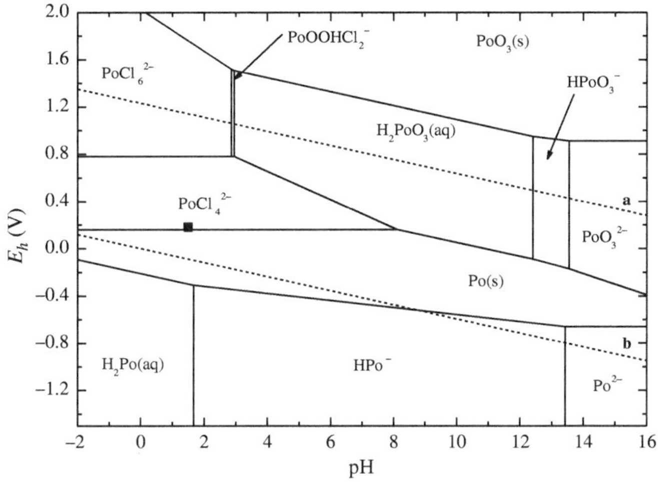

Figure 5.34 pH–potential diagram for the polonium–chloride–water system [(Po) = 10^{-12} mol L^{-1}; (Cl) = 0.12 mol L^{-1}].

In the presence of silver metal, any polonium in solution will react with the silver to produce the very stable silver polonide. The equation for this reaction is:

$$2Ag(s) + PoCl_4^{2-} \rightleftharpoons Ag_2Po(s) + 4Cl^- \qquad (5.10)$$

and the Gibbs energy for the reaction is -158.2 kJ mol^{-1} (equivalent to a reaction stability constant of log $K = 27.7$, thus demonstrating the high stability). Fig. 5.35 demonstrates the speciation of polonium in the presence of silver [in the figure, the concentrations used are 10^{-12} mol L^{-1} for polonium and 10^{-10} mol L^{-1} for silver; however, the latter concentration is unimportant since the relevant silver species are all solid phases (as indicated by Eq. 5.9)].

Figs. 5.34 and 5.35 show the processes that occur in the autodeposition of polonium onto silver. In the final step, the deposition occurs due to the substantial stability of silver polonide that will result in the sequestration of any polonium in solution onto, and within, the silver plate. Elevated temperature is required in the iron dissolution step to ensure that the polonium remains in solution. As is evident in Fig. 5.34 (which is created for a temperature of 25°C; there are insufficient data to construct a pH–potential diagram for elevated temperature), the polonium speciation is close to the $PoCl_4{}^{2-}$–elemental polonium boundary, meaning that at this lower temperature there is potential to lose any polonium that occurs

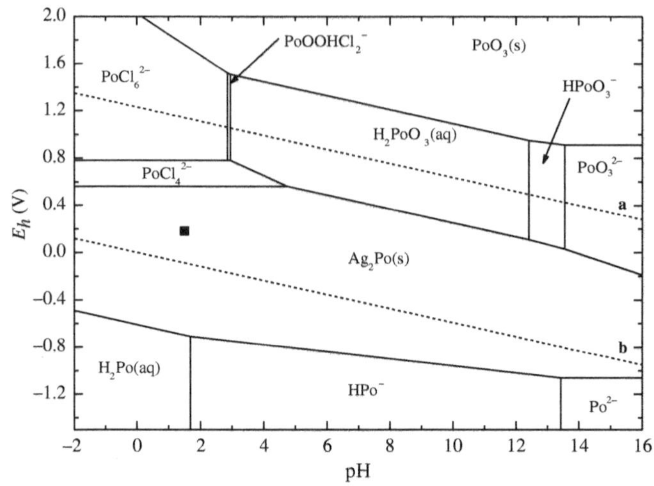

Figure 5.35 pH–potential diagram for the polonium–silver–chloride–water system [(Po) = 10^{-12} mol L^{-1}; (Ag) = 10^{-10} mol L^{-1}; (Cl) = 0.12 mol L^{-1}].

in elemental form with impurities removed in this step (e.g., elemental selenium). Elevated temperature favors the formation of $PoCl_4^{2-}$ over elemental polonium.

5.8 Case study—the mineral processing of rare earth minerals

The mineral processing of rare earth elements (REEs) is necessarily ore specific and, as a result, process flowsheets are quite varied, being tailored to the gangue minerals present and the nature of the host ore (i.e., contain monazite or bastnasite) (Haque et al., 2014). Nevertheless, the most commonly employed processing methods involve either caustic conversion or sulfuric acid decomposition. Of these two methods, caustic conversion is more suitable for industrial application (Anvia, 2015). Caustic conversion involves fewer steps and employs a lower reaction temperature.

Approximately 95% of REE deposits occur with only three minerals: monazite, xenotime, and bastnasite. These minerals can contain natural radioactivity, both from the thorium and uranium decay chains. Monazite can have a thorium content of 4%–12% and a uranium content up to 0.5%, whereas in bastnasite both the thorium and uranium contents are typically much lower (Anvia, 2015). Therefore it is important to understand the radionuclide deportment in the processing of these minerals, particularly monazite.

The processing of monazite using the caustic conversion route uses three steps: caustic conversion, separation of thorium from the REEs using a mild acidic chloride leach, and radioactive impurity removal. Anvia (2015) conducted a detailed study of REE processing of monazite using this three step process and documented the deportment of the radionuclides, including polonium. This study will be used here to further define the chemical behavior of both polonium and lead in the processing of monazite.

Anvia (2015) studied a monazite concentrate that contained 51% REEs (including yttrium) that was dominated by the light REEs, lanthanum, cerium, praseodymium, and neodymium. The sample also contained 6% thorium and 0.22% uranium. The mineralogy of the concentrate was predominantly monazite (94%), with lesser amounts of zircon (4.4%), clay minerals (0.4%), xenotime (0.2%), and others (0.6%). The concentrate had a ^{238}U concentration of 32 Bq g^{-1} and a ^{210}Pb and ^{210}Po concentration of 27 Bq g^{-1}.

Monazite contains the REEs and thorium as phosphates. The caustic conversion process, often called cracking, converts the phosphates into hydroxides. The process uses concentrated sodium hydroxide and, as such, the pH is about 15. Unfortunately, Anvia (2015) did not measure the redox potential (E_h) in any of the process steps. Consequently, herein the E_h has been inferred from the behavior of the radionuclides, polonium, and lead. In the caustic conversion step, the deportment of nonradioactive lead and ^{210}Pb was found to be the same, even though these two forms of lead may be sited differently within the monazite matrix, with 25% and 31%, respectively, being solubilized in the sodium phosphate liquor. This similarity suggests that both nonradioactive lead and ^{210}Pb undergo the same reactions. At the pH used in the caustic conversion process, lead is largely soluble as $Pb(OH)_3^-$ (see Fig. 5.36). Thus a significant proportion of lead remaining in the solid suggests either a low redox potential in the cracking process or that both nonradioactive lead and ^{210}Pb are associated with refractory minerals. Of these two possibilities, the latter does not seem likely due to the similar behavior of both ^{210}Pb and nonradioactive lead and, as such, an E_h of about -0.75 V has been assigned for the caustic cracking step. This E_h is assigned as a consequence of the line of equilibrium shown in Fig. 5.36 between $Pb(OH)_3^-$ and metallic lead. For lead to be predominantly insoluble in the caustic conversion step, the E_h should lie slightly below this line of equilibrium. Another possibility is that the E_h is somewhat higher, but lead occurs

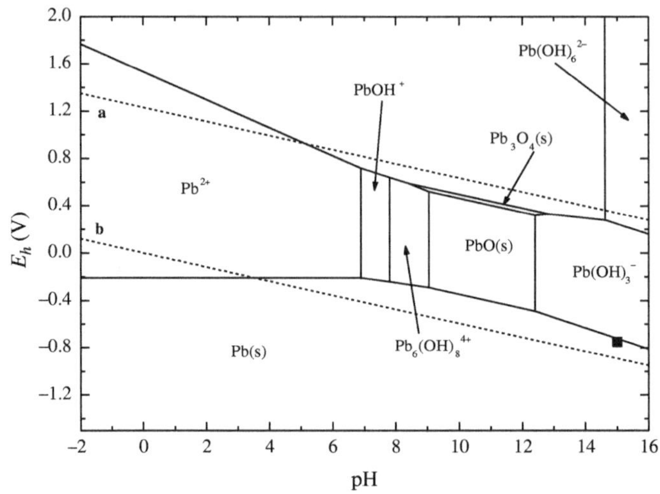

Figure 5.36 pH–potential diagram for the lead–water system in caustic conversion [(Pb) = 0.0025 mol L^{-1}; pH = 15; $E_h = -0.75$ V].

mostly in solid form either as $NaPb(OH)_3$(s) or Na_2PbO_2(s). The latter phase is well known and has been used extensively in the petrochemical industry for the removal or analysis of mercaptans in petroleum products (de Angelis, 2012).

The caustic conversion of different monazite samples may lead to differing final conditions (e.g., in pH and E_h) as a result of differences in the composition of the monazite ore. Hart and Levins (1988) found that only 19% of ^{210}Pb was found in the sodium phosphate liquor. Although this result is in reasonable agreement with the study of Anvia (2015), it suggests that either the process pH or E_h was marginally lower in the study of Hart and Levins (1988).

The behavior of polonium can be projected from the conditions (pH = 15 and $E_h = -0.75$) assigned for lead. At these conditions, polonium will be present as the polonide ion, Po^{2-} (see Fig. 5.37). This might suggest that polonium should be soluble. However, in the presence of a much larger amount of elemental lead, the polonide ion would react to form insoluble lead polonide, in a similar manner to that described by Eq. (5.3), except where the reacting species is Po^{2-} rather than HPo^-. If, however, the E_h was higher than shown in Fig. 5.37 and lead precipitates as a sodium plumbite phase, then for polonium to remain insoluble (as either an insoluble polonide or polonium metal) the E_h would need to be less than about -0.4 V.

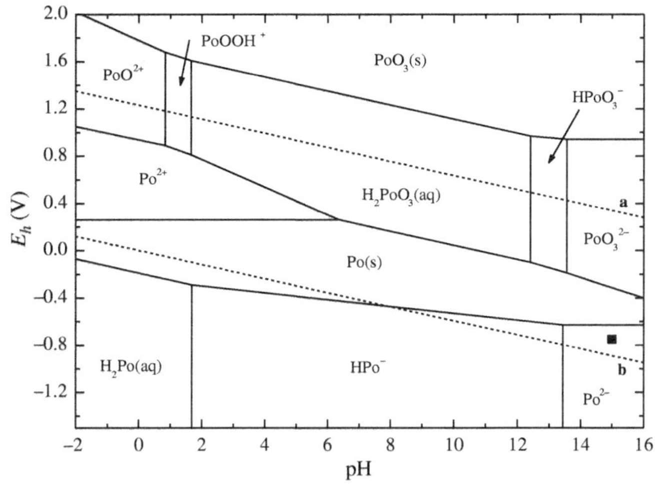

Figure 5.37 pH–potential diagram for the polonium–water system in caustic conversion [(Po) = 1.67×10^{-13} mol L^{-1}; pH = 15; $E_h = -0.75$ V].

In the processing of monazite, the caustic conversion step transforms the rare earth and thorium phosphates into insoluble hydroxide phases. Thorium is separated from the rare earths by leaching with hydrochloric acid at a pH of 3.2 (Anvia, 2015). Within this step, the REE will be converted into soluble chlorides whereas thorium will remain as an insoluble hydroxide phase with the final REE concentration being 160 g L^{-1}, from which a chloride concentration of 3.42 mol L^{-1} is derived.

Typically, polonium is soluble in concentrated chloride solutions; however, Anvia (2015) showed that only 6.3% of the polonium was solubilized in the hydrochloric acid leach step. In this step, it would appear that there are few, if any, reactions that will affect the redox behavior of the solution. As such, it might be expected that (pH + pe) will be a constant [i.e., (pH + pe) is equivalent following both caustic conversion and hydrochloric acid leaching]. In the caustic conversion step, the assigned pH was 15 and the E_h was -0.75 V (=pe of -12.66), leading to a value of (pH + pe) of 2.34. The pH of the hydrochloric acid leaching step is 3.2 which, based on an equivalence of (pH + pe), leads to a pe of -0.86. This is equivalent to an E_h of -0.051 V. If this latter E_h is assumed, it is expected that polonium will be predominantly insoluble (Fig. 5.38) during the hydrochloric acid leaching step and will be present predominantly as polonium metal. This predicted insolubility is consistent with the findings of Anvia (2015).

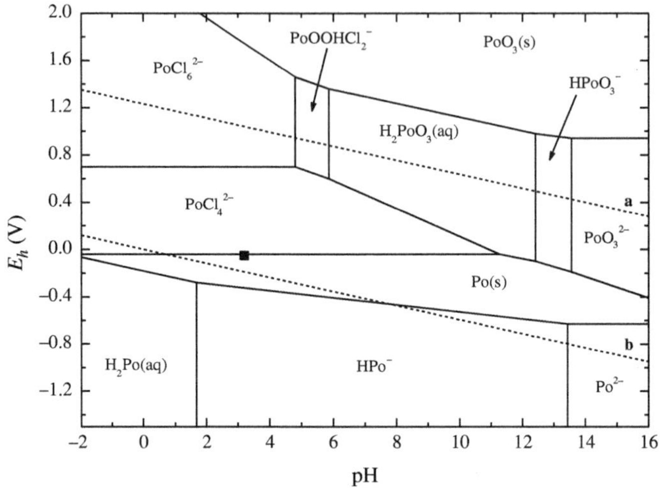

Figure 5.38 pH–potential diagram for the polonium–chlorine–water system in hydrochloric acid leaching [(Po) = 1.09×10^{-13} mol L^{-1}; (Cl^-) = 3.42 mol L^{-1}; pH = 3.2; $E_h = -0.051$ V].

In the hydrochloric acid leach, Anvia (2015) found that 43% of lead was solubilized into the chloride solution and 48% of ^{210}Pb. Seward (1984) studied the complexation of lead in chloride solutions, up to a chloride concentration of 3.2 mol kg^{-1}. These data indicate that at elevated chloride concentrations, five lead species will occur in solution, Pb^{2+}, $PbCl^+$, $PbCl_2$(aq), $PbCl_3^-$, and $PbCl_4^{2-}$, with the latter species being dominant. In the presence of elevated chloride concentrations, lead can also form a solid phase, $PbCl_2$(s) (cotunnite). Utilizing the lead chloride stability constants derived from Seward (1984) and the solubility constant of cotunnite, speciation calculations indicate that the solid phase dominates (50.4%), with all four lead–chloride complexes accounting for various fractions of the soluble lead (a total of 49.6%). The predicted concentration of lead in solution is in very good agreement with that observed by Anvia (2015). The pH–potential diagram for the lead–chloride–water system is illustrated in Fig. 5.39.

In the final step, radionuclides are separated from REE by precipitation using either sulfate (radium) or sulfide (lead and polonium). After the hydrochloric acid leach step, just over 30% of the ^{210}Pb remains in the REE chloride solution and about 6% of the ^{210}Po. Anvia (2015) used 0.5, 2.5, and 5.0 g L^{-1} of sulfide ion (added as sodium sulfide) to effect removal of, in particular, lead and found that 0.5 g L^{-1} was optimum. Lead sulfide is very insoluble in aqueous solutions.

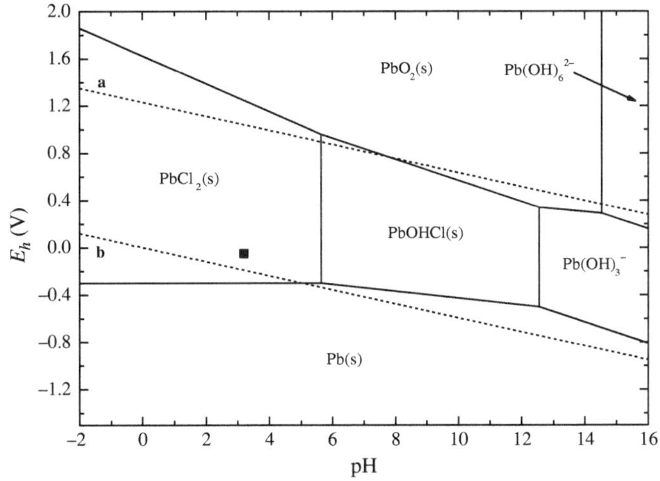

Figure 5.39 pH–potential diagram for the lead–chlorine–water system in hydrochloric acid leaching [(Pb) = 0.0018 mol L^{-1}; (Cl^-) = 3.42 mol L^{-1}; pH = 3.2; E_h = −0.051 V].

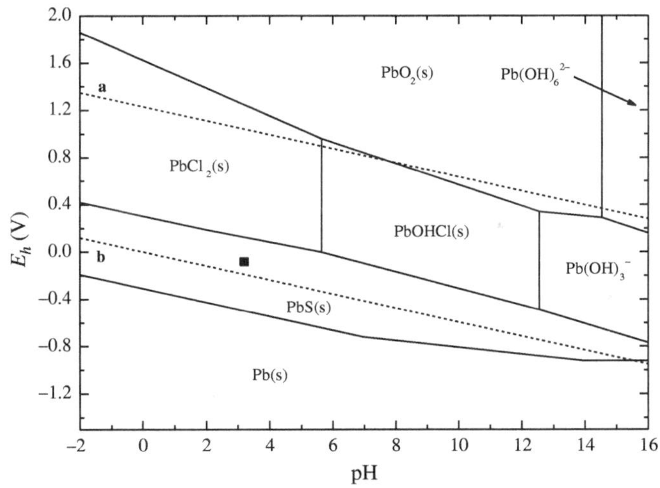

Figure 5.40 pH–potential diagram for the lead–chlorine–sulfur–water system in radionuclide removal [(Pb) = 0.0018 mol L^{-1}; (Cl^-) = 3.42 mol L^{-1}; (S) = 0.016 mol L^{-1}; pH = 3.2; E_h = −0.082 V].

The introduction of sodium sulfide into the hydrochloric acid leach liquor will induce a decrease in E_h, but with an addition of only 0.5 g L^{-1} sulfide the reduction in E_h will be relatively minor. However, the reduction in E_h for the three concentrations studied by Anvia (2015) will all result in solid lead sulfide being the predominant lead species. Fig. 5.40 illustrates the pH–potential behavior of lead (pH = 3.2 and

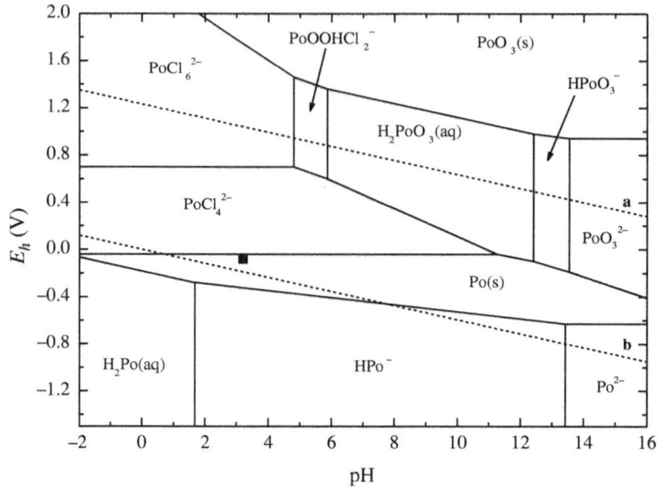

Figure 5.41 pH–potential diagram for the polonium–chlorine–sulfur–water system in radionuclide removal [(Po) = 1.09×10^{-13} mol L^{-1}; (Cl^-) = 3.42 mol L^{-1}; (S) = 0.016 mol L^{-1}; pH = 3.2; $E_h = -0.082$ V].

$E_h = -0.082$ V) in the presence of both chloride (3.42 mol L^{-1}) and sulfur (0.016 mol L^{-1}).

The reduction in E_h induces further polonium to be removed from solution. Even though the change in E_h from the sulfide addition is relatively small, the reduction is sufficient for removal of the majority of the remaining polonium from the chloride liquor (see Fig. 5.41). Therefore the addition of sodium sulfide is capable of removing both ^{210}Pb and ^{210}Po from the REE chloride liquor.

References

Ahrland, S., 1975. Metal complexes present in seawater. In: Goldberg, E.D. (Ed.), The Nature of Seawater, Dahlem Workshop Report. Dr S. Bernhard, Dahlem Konferenzen, Berlin, pp. 219–220.

Amer, A.M., 2002. Processing of copper anode-slimes for extraction of metal values. Phys. Prob. Min. Proc 36, 123–134.

Ansoborlo, E., Berard, P., Den Auwer, C., Leggett, R., Menetrier, F., Younes, A., et al., 2012. Review of chemical and radiotoxicological properties of polonium for internal contamination purposes. Chem. Res. Toxicol. 25, 1551–1564.

Anvia, M., 2015. Radionuclide Deportment in Rare Earth Processing From Monazite and Bastnasite Using Conventional and Alternative Processing Routes (Ph.D. dissertation). University of Sydney.

Bagnall, K.W., 1957a. The chemistry of polonium. Quart. Rev. (London) 11, 30–48.

Biswas, A.K., Davenport, W.G., 1994. Extractive Metallurgy of Copper., third ed. Pergamon Press, pp. 330–332.

Bockris, J.O.M., Reddy, A.K.N., 1970. Modern Electrochemistry, vol. 2. Plenum Press, New York, pp. 1281–1285.
Brown, S.A., 1998. Behaviour of polonium-210 during anode slimes processing. In: Proceedings of SPERA98, Radioactivity and the Environment. Christchurch, New Zealand. pp. 102–107. Available from: https://inis.iaea.org/collection/NCLCollectionStore/_Public/30/057/30057784.pdf.
Brown, S.A., 2001. The Aqueous Chemistry of Polonium and Its Relationship to Mineral Processing Streams (Ph.D. dissertation). University of Western Sydney.
de Angelis, A., 2012. Natural gas removal of hydrogen sulfide and mercaptans. Appl. Catalysis B: Environ 113-114, 37–42.
Dimitrijević, S.P., Anðić, Z., Kamberović, Ž., Dimitrijević, S.B., Vuković, N., 2014. Recycling of silver-plated brass for production of high purity copper and ultrafine silver powder for electric contacts. Bulg. Chem. Comm 46, 814–824.
Garrels, R.M., Christ, C.L., 1965. Solutions, Minerals and Equilibria. Cooper and Company, San Francisco, Freeman, pp. 42–43.
Grayson, M. (Ed.), 1978. Kirk-Othmer Encyclopedia of Chemical Technology, third ed. vol. 4. John Wiley & Sons Inc, New York, p. 561.
Hall, S., 1993. Gold and silver recovery from copper anode slimes at The Olympic Dam Joint Venture, Roxby Downs, SA. In: second ed. Woodcock, J.T., Hamilton, J.K. (Eds.), Australasian Mining and Metallurgy – The Sir Maurice Mawby Memorial Volume, vol. 2. The Australasian Institute of Mining and Metallurgy, Parkville, pp. 1102–1105.
Haque, N., Hughes, A., Lim, S., Vernon, C., 2014. Rare earth elements: overview of mining, mineralogy, uses, sustainability and environmental impact. Resources 3, 614–635.
Hart, K.P., Levins, D.M., 1988. Management of wastes from the processing of rare earth minerals. Chemeca 88. Sydney, Australia.
He, S., Wang, J., Xu, Z., Wang, J., Gan, L., 2014. Removal of copper and enrichment of precious metals by pressure leaching pretreatment of copper anode slimes in sulfuric acid medium. Precious Metals 35, 48–53.
Kadirvelu, K., Faur-Brasquet, C., Le Cloirec, P., 2000. Removal of Cu(II), Pb(II) and Ni (II) by adsorption onto activated carbon cloths. Langmuir 16, 8404–8409.
Kester, D.R., Ahrland, S., Beasley, T.M., Bernhard, M., Branica, M., Campbell, I.D., et al., 1975. Chemical speciation in seawater – group report. In: Goldberg, E.D. (Ed.), The Nature of Seawater, Dahlem Workshop Report. Dr S. Bernhard, Dahlem Konferenzen, Berlin, pp. 17–41.
Kocher, D.C., 1977. Nuclear Decay Data for Radionuclides Occurring in Routine Releases From Nuclear Fuel Cycle Facilities. Oak Ridge National Laboratory, ORNL/NUREG/T102.
Kolthoff, I.M., Elving, P.J. (Eds.), 1961a. Treatise on Analytical Chemistry: Part II, vol. 7. John Wiley & Sons, Inc, New York, p. 140.
Kolthoff, I.M., Elving, P.J. (Eds.), 1961b. Treatise on Analytical Chemistry: Part II, vol. 7. John Wiley & Sons, Inc, New York, p. 150.
Kolthoff, I.M., Elving, P.J. (Eds.), 1964. Treatise on Analytical Chemistry: Part II, vol. 6. John Wiley & Sons, Inc, New York, p. 527.
Kroschwitz, J.I. (Ed.), 1993. Kirk-Othmer Encyclopedia of Chemical Technology, fourth ed. vol. 7. John Wiley & Sons Inc, New York, pp. 381–428.
Kroschwitz, J.I. (Ed.), 1994. Kirk-Othmer Encyclopedia of Chemical Technology, fourth ed. vol. 12. John Wiley & Sons, Inc, New York, p. 741.
Kroschwitz, J.I. (Ed.), 1995. Kirk-Othmer Encyclopedia of Chemical Technology, fourth ed. vol. 16. John Wiley & Sons, Inc, New York, p. 351.

Kroschwitz, J.I. (Ed.), 1997. Kirk-Othmer Encyclopedia of Chemical Technology, fourth ed. vol. 21. John Wiley & Sons, Inc, New York, p. 688.
Leigh, A.H., 1981. The recovery of precious metals from tank house slime. Symposium on Recovery, Reclamation and Refining of Precious Metals. San Diego, CA.
Lu, D., Chang, Y., Yang, H., Feng, X., 2015. Sequential removal of selenium and tellurium from copper anode slime with high nickel content. Trans. Nonferrous Met. Soc. China 25, 1307–1314.
Lyklema, J., 1975. Interfacial electrochemistry of hydrophobic colloids. In: Goldberg, E.D. (Ed.), The Nature of Seawater, Dahlem Workshop Report. Dr S. Bernhard, Dahlem Konferenzen, Berlin, pp. 579–586.
Marsden, J.O., House, C.I., 2006. The Chemistry of Gold Extraction. Society for Mining, Metallurgy and Exploration, second ed. Littleton.
Peck, G.A., Smith, J.D., 2000. Distribution of dissolved and particulate ^{226}Ra, ^{210}Pb and ^{210}Po in the Bismark Sea and western equatorial Pacific Ocean. Mar. Freshwat. Res. 51, 647–658.
Saeedi, M., Alamdari, E.K., Fatmehsari, D.H., Darvishi, D., Alamdari, A.K., kafash Rajsanjami, A.B., 2013. Effect of Temperature on Dissolution of Gold from Copper Anode Slime in 23rd International Mining Congress and Exhibition of Turkey. Antalya, pp. 1347–1351.
Seward, T.M., 1984. The formation of lead(II) chloride complexes to 300 °C: a spectrophotometric study. Geochim. Cosmochim. Acta 48, 121–134.
Shannon, R.D., 1976. Revised effective ionic radii and systematic studies of interatomic distances in halides and chalcogenides. Acta Cryst. A32, 751–767.
Sienko, M.J., Plane, R.A., 1976. Chemistry, fifth ed. McGrw-Hill Kogakusha Ltd, Tokyo, p. 200.
Sillén, L.G., Martell, A.E., 1964. Stability Constants of Metal-Ion Complexes. The Chemical Society, London, Special Publication No. 17.
Skoog, D.A., West, D.M., 1982. Fundamentals of Analytical Chemistry, 4th Edition, CBS College Publishing, Philadelphia, p. 132.
Smith, A., Mudder, T., 1991. The Chemistry and Treatment of Cyanide Wastes. Mining Journal Books Ltd, p. 6.
Stanley, G.G. (Ed.), 1987. The Extractive Metallurgy of Gold in South Africa. National Book Printers, Goodwood, p. 895.
Towler, P.H., Smith, J.D., 1997. Distribution of ^{226}Ra and ^{210}Pb in the mixed layer of the western equatorial Pacific Ocean. Mar. Freshwat. Res. 48, 371–375.
Vanýek, P., 1996. Electrochemical Series. In: Lide, D.R. (Ed.), CRC Handbook of Chemistry and Physics, 77th Edition, CRC Press Inc, Boca Raton, pp. 8–22.
Vogel, A.I., 1961a. A Textbook of Quantitative Inorganic Analysis., third ed. Longmans, London, pp. 271–272.
Vogel, A.I., 1961b. A Textbook of Quantitative Inorganic Analysis., third ed. Longmans, London, pp. 483–484.
Vogel, A.I., 1961c. A Textbook of Quantitative Inorganic Analysis., third ed. Longmans, London, pp. 110–111.
Vonk, A.S.M., 1993. Copper concentrator practice at the Olympic Dam Joint Venture, Roxby Downs, SA. In: second ed. Woodcock, J.T., Hamilton, J.K. (Eds.), Australasian Mining and Metallurgy – The Sir Maurice Mawby Memorial Volume, vol. 1. The Australasian Institute of Mining and Metallurgy, Parkville, pp. 656–658.
Yannopoulos, J.C., 1991. The Extractive Metallurgy of Gold. Van Nostrand Reinhold, New York.

CHAPTER 6

Conclusions

It has been somewhat surprising to find from the literature survey that most of the experimental work to determine the characteristics and properties of polonium was conducted prior to 1960. One would assume that advances in technology would make such studies easier to pursue and yet our knowledge of the chemistry of polonium, as compared with other elements, is scanty. Nevertheless it has been possible, through a critical evaluation of the data scattered in the literature, to elucidate the basics of the chemical properties of the element and present this as a thermochemical database. In providing such a tool for use within the scientific community, it is hoped that published experimental results, and indeed those that may lay buried because of a lack of understanding, will be reassessed in light of this newly acquired knowledge to further clarify the aqueous chemistry of this element.

Polonium was found to have characteristics both of lead and tellurium, depending upon conditions. In a pure water system, under acidic conditions, its behavior is similar to that of lead because of its ability to form the divalent cation, Po^{2+}. Under basic conditions, however, its chemistry is similar to that of tellurium (and selenium) with aqueous species dominant for these elements in this region.

Processing of ores containing polonium is one area that has highlighted the difficulties in trying to remove/control polonium in plant circuits because of a lack of understanding of its solution chemistry. The derivation of pH-potential diagrams for these circuits from the acquired thermochemical database represents a successful attempt to link theory with practical measurements, to model observed behavior. Surprisingly, the behavior of polonium could be explained in terms of these diagrams, and with respect to lead, tellurium and selenium, in all but one case; that under the conditions of sulfuric acid leaching. Here, polonium was found to remain as a solid and not to dissolve as predicted by the diagrams. Volatilized polonium is, however, known to be leached by sulfuric acid from smelter dusts (Ring et al., 1995) and so the conclusions made regarding this nonleachability being due to some form of siting phenomena within the matrix, based on the chemistry of upstream processing, seem valid. It is also plausible that the polonium coprecipitates with lead as a sulfate phase.

The Aqueous Chemistry of Polonium and the Practical Application of its Thermochemistry
DOI: https://doi.org/10.1016/B978-0-12-819308-2.00009-7

Overall this work has gathered together the current knowledge of polonium chemistry to provide a coherent explanation of its solution chemistry. It has been shown that the approach adopted can be applied to real systems involving the element. The correlation between predicted behavior, for several case studies, and practical observations of polonium chemistry have been, in general, very good, and hence validate the derived thermochemical database values. Further refinements of the model await careful experimental work that is both heroic and taxing, in the extreme.

Reference

Ring, R.J., Collier, D.E., Day, R.A., 1995. Leaching of smelter dust to remove polonium. Report to WMC—Olympic Dam Operations, ANSTO/C432.

APPENDIX 1

Thermochemical data

Selected thermochemical data at 101 kPa and 298.15K

Species	Reaction	Parameter	Value	Reference
H_2O	$\frac{1}{2}O_2(g) + 2H^+ \rightleftharpoons H_2O$	ΔG_f°	− 237.17	Brown and Ekberg (2016)
		E°	1.229	Calculated from ΔG_f°
OH^-	$H_2O \rightleftharpoons OH^- + H^+$	ΔG_f°	− 157.3	Brown and Ekberg (2016)
		$\log K^\circ$	− 13.994	
F^-	$\frac{1}{2}F_2(g) + e^- \rightleftharpoons F^-$	ΔG_f°	− 281.5	Grenthe et al. (1992)
		E°	2.918	Calculated from ΔG_f°
Cl^-	$\frac{1}{2}Cl_2(g) + e^- \rightleftharpoons Cl^-$	ΔG_f°	− 131.2	Grenthe et al. (1992)
		E°	1.36	Calculated from ΔG_f°
Br^-	$\frac{1}{2}Br_2(l) + e^- \rightleftharpoons Br^-$	ΔG_f°	− 103.9	Grenthe et al. (1992)
		E°	1.076	Calculated from ΔG_f°
I^-	$\frac{1}{2}I_2(s) + e^- \rightleftharpoons I^-$	ΔG_f°	− 51.7	Grenthe et al. (1992)
		E°	0.536	Calculated from ΔG_f°
S^{2-}	$S(s) + 2e^- \rightleftharpoons S^{2-}$	ΔG_f°	93.2	Calculated from E°
		E°	− 0.483	Sillén and Martell (1964, 1971)
Se^{2-}	$Se(s) + 2e^- \rightleftharpoons Se^{2-}$	ΔG_f°	128.6	Olin et al. (2005)
		E°	− 0.666	Calculated from ΔG_f°
Te^{2-}	$Te(s) + 2e^- \rightleftharpoons Te^{2-}$	ΔG_f°	183.3	Calculated from E°
		E°	− 0.95	Panson (1963)
HTe^-	$H_2Te(aq) \rightleftharpoons HTe^- + H^+$	ΔG_f°	113.5	Calculated from $\log K^\circ$
		$\log K^\circ$	− 2.64	McPhail (1995)
$H_2Te(aq)$	$Te(s) + 2H^+ + 2e^- \rightleftharpoons H_2Te(aq)$	ΔG_f°	98.4	Calculated from E°
		E°	− 0.51	Panson (1963)
$TeO_2(s)$	$Te(s) + O_2(g) \rightleftharpoons TeO_2(s)$	ΔG_f°	− 270.3	Zhdanov (1985)
$TeOOH^+$	$Te(s) + O_2(g) + H^+ \rightleftharpoons TeOOH^+$	ΔG_f°	− 261.54	Zhdanov (1985)
$H_2TeO_3(aq)$	$Te(s) + 1.5O_2(g) + H_2(g) \rightleftharpoons H_2TeO_3(aq)$	ΔG_f°	− 474.6	McPhail (1995)
$HTeO_3^-$	$Te(s) + 2H^+ + 2e^- \rightleftharpoons H_2Te(aq)$	ΔG_f°	98.4	Calculated from E°
		E°	− 0.51	Panson (1963)
TeO_3^{2-}	$Te(s) + 1.5O_2(g) + 2e^- \rightleftharpoons TeO_3^{2-}$	ΔG_f°	− 384	McPhail (1995)
$H_2S(aq)$	$S(s) + 2H^+ + 2e^- \rightleftharpoons H_2S(aq)$	ΔG_f°	− 27.6	Grenthe et al. (1992)
		E°	0.143	Calculated from ΔG_f°
$H_2Se(aq)$	$Se(s) + 2H^+ + 2e^- \rightleftharpoons H_2Se(aq)$	ΔG_f°	21.5	Olin et al. (2005)
		E°	− 0.111	Calculated from ΔG_f°

HS^-	$H_2S(aq) \rightleftharpoons HS^- + H^+$	ΔG_f°	12.2	Grenthe et al. (1992)
		$\log K^\circ$	− 6.99	Calculated from ΔG_f°
HSe^-	$H_2Se(aq) \rightleftharpoons HSe^- + H^+$	ΔG_f°	43.5	Olin et al. (2005)
		$\log K^\circ$	− 3.85	Calculated from ΔG_f°
$Cl_2(g)$		S_f°	223.08	Grenthe et al. (1992)
$InCl(s)$	$In(s) + \frac{1}{2}Cl_2(g) \rightleftharpoons InCl(s)$	ΔH_f°	− 186	Bard et al. (1985)
$InBr(s)$	$In(s) + \frac{1}{2}Br_2(l) \rightleftharpoons InBr(s)$	ΔG_f°	− 169	Bard et al. (1985)
		ΔH_f°	− 175	
$InI(s)$	$In(s) + \frac{1}{2}I_2(s) \rightleftharpoons InI(s)$	ΔG_f°	− 120	Bard et al. (1985)
		ΔH_f°	− 116	
$TlCl(s)$	$Tl(s) + \frac{1}{2}Cl_2(g) \rightleftharpoons TlCl(s)$	ΔH_f°	− 204.1	Bard et al. (1985)
$TlBr(s)$	$Tl(s) + \frac{1}{2}Br_2(l) \rightleftharpoons TlBr(s)$	ΔG_f°	− 167	Bard et al. (1985)
		ΔH_f°	− 173	
$TlI(s)$	$Tl(s) + \frac{1}{2}I_2(s) \rightleftharpoons TlI(s)$	ΔG_f°	− 125.4	Bard et al. (1985)
		ΔH_f°	− 123.8	
$SbCl_3(s)$	$Sb(s) + 1.5Cl_2(g) \rightleftharpoons SbCl_3(s)$	ΔG_f°	− 323.7	Bard et al. (1985)
		ΔH_f°	− 382.3	
$SbBr_3(s)$	$Sb(s) + 1.5Br_2(l) \rightleftharpoons SbBr_3(s)$	ΔG_f°	− 239.3	Bard et al. (1985)
		ΔH_f°	− 259.4	
$BiCl_3(s)$	$Bi(s) + 1.5Cl_2(g) \rightleftharpoons BiCl_3(s)$	ΔG_f°	− 314.6	Bard et al. (1985)
		ΔH_f°	− 379.1	
$BiBr_3(s)$	$Bi(s) + 1.5Br_2(l) \rightleftharpoons BiBr_3(s)$	ΔG_f°	− 247.7	Bard et al. (1985)
		ΔH_f°	− 276.1	
$SnCl_2(s)$	$Sn(s) + Cl_2(g) \rightleftharpoons SnCl_2(s)$	ΔG_f°	− 302.1	Bard et al. (1985)
		ΔH_f°	− 349.8	
$SnBr_2(s)$	$Sn(s) + Br_2(l) \rightleftharpoons SnBr_2(s)$	ΔG_f°	− 248.9	Bard et al. (1985)
		ΔH_f°	− 266.1	
$PbCl_2(s)$	$Pb(s) + Cl_2(g) \rightleftharpoons PbCl_2(s)$	ΔG_f°	− 314	Bard et al. (1985)
		ΔH_f°	− 359.2	
$PbBr_2(s)$	$Pb(s) + Br_2(l) \rightleftharpoons PbBr_2(s)$	ΔG_f°	− 260.4	Bard et al. (1985)
		ΔH_f°	− 277	
$TeCl_6^{2-}$	$Te(s) + 3Cl_2(g) + 2e^- \rightleftharpoons TeCl_6^{2-}$	ΔG_f°	− 574.9	Bard et al. (1985)

(*Continued*)

(Continued)

Species	Reaction	Parameter	Value	Reference
$TeCl_4(s)$	$Te(s) + 2Cl_2(g) \rightleftharpoons TeCl_4(s)$	ΔG_f^o	− 237.2	Bard et al. (1985)
		ΔH_f^o	− 326.4	
$AgCl(aq)$	$Ag^+ + Cl^- \rightleftharpoons AgCl(aq)$	$\log K^o$	3.18	Sillén and Martell (1964)
$AgCl_2^-$	$Ag^+ + 2Cl^- \rightleftharpoons AgCl_2^-$	$\log K^o$	5.18	Sillén and Martell (1964)
$BiCl^{2+}$	$Bi^{3+} + Cl^- \rightleftharpoons BiCl^{2+}$	$\log K^o$	3.65	Lothenbach et al. (1999)
$BiCl_2^+$	$Bi^{3+} + 2Cl^- \rightleftharpoons BiCl_2^+$	$\log K^o$	5.85	Lothenbach et al. (1999)
$BiCl_3(aq)$	$Bi^{3+} + 3Cl^- \rightleftharpoons BiCl_3(aq)$	$\log K^o$	7.62	Lothenbach et al. (1999)
$BiCl_4^-$	$Bi^{3+} + 4Cl^- \rightleftharpoons BiCl_4^-$	$\log K^o$	9.06	Lothenbach et al. (1999)
$CdCl^+$	$Cd^{2+} + Cl^- \rightleftharpoons CdCl^+$	$\log K^o$	1.98	Powell et al. (2011)
$CdCl_2(aq)$	$Cd^{2+} + 2Cl^- \rightleftharpoons CdCl_2(aq)$	$\log K^o$	2.64	Powell et al. (2011)
$CdCl_3^-$	$Cd^{2+} + 3Cl^- \rightleftharpoons CdCl_3^-$	$\log K^o$	2.3	Powell et al. (2011)
$CdCl_4^{2-}$	$Cd^{2+} + 4Cl^- \rightleftharpoons CdCl_4^{2-}$	$\log K^o$	1.6	Sillén and Martell (1964)
$FeCl^{2+}$	$Fe^{3+} + Cl^- \rightleftharpoons FeCl^{2+}$	$\log K^o$	1.52	Lemire et al. (2013)
$FeCl_2^+$	$Fe^{3+} + 2Cl^- \rightleftharpoons FeCl_2^+$	$\log K^o$	2.22	Lemire et al. (2013)
$FeCl_3(aq)$	$Fe^{3+} + 3Cl^- \rightleftharpoons FeCl_3(aq)$	$\log K^o$	1.02	Lemire et al. (2013)
$CuCl^+$	$Cu^{2+} + Cl^- \rightleftharpoons CuCl^+$	$\log K^o$	0.83	Powell et al. (2007)
$CuCl_2(aq)$	$Cu^{2+} + 2Cl^- \rightleftharpoons CuCl_2(aq)$	$\log K^o$	0.6	Powell et al. (2007)
$PbCl^+$	$Pb^{2+} + Cl^- \rightleftharpoons PbCl^+$	$\log K^o$	1.5	Powell et al. (2009)
$PbCl_2(aq)$	$Pb^{2+} + 2Cl^- \rightleftharpoons PbCl_2(aq)$	$\log K^o$	2.1	Powell et al. (2009)
$PbCl_3^-$	$Pb^{2+} + 3Cl^- \rightleftharpoons PbCl_3^-$	$\log K^o$	2	Powell et al. (2009)
$PbCl_4^{2-}$	$Pb^{2+} + 4Cl^- \rightleftharpoons PbCl_4^{2-}$	$\log K^o$	1.37	Sillén and Martell (1964)
$PdCl^+$	$Pd^{2+} + Cl^- \leftrightarrow PdCl^+$	$\log K^o$	6.1	Sillén and Martell (1964)
$PdCl_2(aq)$	$Pd^{2+} + 2Cl^- \rightleftharpoons PdCl_2(aq)$	$\log K^o$	10.7	Sillén and Martell (1964)
$PdCl_3^-$	$Pd^{2+} + 3Cl^- \rightleftharpoons PdCl_3^-$	$\log K^o$	13.1	Sillén and Martell (1964)
$PdCl_4^{2-}$	$Pd^{2+} + 4Cl^- \rightleftharpoons PdCl_4^{2-}$	$\log K^o$	15.7	Sillén and Martell (1964)
$SnCl^+$	$Sn^{2+} + Cl^- \rightleftharpoons SnCl^+$	$\log K^o$	1.52	Gamsjäger et al. (2012)
$SnCl_2(aq)$	$Sn^{2+} + 2Cl^- \rightleftharpoons SnCl_2(aq)$	$\log K^o$	2.17	Gamsjäger et al. (2012)
$SnCl_3^-$	$Sn^{2+} + 3Cl^- \rightleftharpoons SnCl_3^-$	$\log K^o$	2.13	Gamsjäger et al. (2012)
$SnCl_4^{2-}$	$Sn^{2+} + 4Cl^- \rightleftharpoons SnCl_4^{2-}$	$\log K^o$	2.03	Gamsjäger et al. (2012)
$SnCl^{3+}$	$Sn^{4+} + Cl^- \rightleftharpoons SnCl^{3+}$	$\log K^o$	3.19	Gamsjäger et al. (2012)
$SnCl_2^{2+}$	$Sn^{4+} + 2Cl^- \rightleftharpoons SnCl_2^{2+}$	$\log K^o$	5.95	Gamsjäger et al. (2012)

$TlCl(aq)$	$Tl^{+} + Cl^{-} \rightleftharpoons TlCl(aq)$	$\log K^{\circ}$	0.6	Sillén and Martell (1964)
$TlCl_2^{-}$	$Tl^{+} + 2Cl^{-} \rightleftharpoons TlCl_2^{-}$	$\log K^{\circ}$	0.18	Sillén and Martell (1964, 1971)
$TlCl^{2+}$	$Tl^{3+} + Cl^{-} \rightleftharpoons TlCl^{2+}$	$\log K^{\circ}$	7.93	Sillén and Martell (1964, 1971)
$TlCl_2^{+}$	$Tl^{3+} + 2Cl^{-} \rightleftharpoons TlCl_2^{+}$	$\log K^{\circ}$	13.54	Sillén and Martell (1964, 1971)
$TlCl_3(aq)$	$Tl^{3+} + 3Cl^{-} \rightleftharpoons TlCl_3(aq)$	$\log K^{\circ}$	16.13	Sillén and Martell (1964, 1971)
$TlCl_4^{-}$	$Tl^{3+} + 4Cl^{-} \rightleftharpoons TlCl_4^{-}$	$\log K^{\circ}$	18.15	Sillén and Martell (1964, 1971)
UO_2Cl^{+}	$UO_2^{2+} + Cl^{-} \rightleftharpoons UO_2Cl^{+}$	$\log K^{\circ}$	0.17	Guillaumont et al. (2003)
$UO_2Cl_2(aq)$	$UO_2^{2+} + 2Cl^{-} \rightleftharpoons UO_2Cl_2(aq)$	$\log K^{\circ}$	− 1.1	Guillaumont et al. (2003)
$ZnCl^{+}$	$Zn^{2+} + Cl^{-} \rightleftharpoons ZnCl^{+}$	$\log K^{\circ}$	0.4	Powell et al. (2013)
$ZnCl_2(aq)$	$Zn^{2+} + 2Cl^{-} \rightleftharpoons ZnCl_2(aq)$	$\log K^{\circ}$	0.69	Powell et al. (2013)
$ZnCl_3^{-}$	$Zn^{2+} + 3Cl^{-} \rightleftharpoons ZnCl_3^{-}$	$\log K^{\circ}$	0.48	Powell et al. (2013)
$ZnCl_4^{2-}$	$Zn^{2+} + 4Cl^{-} \rightleftharpoons ZnCl_4^{2-}$	$\log K^{\circ}$	0.2	Sillén and Martell (1964)
$HgCl^{+}$	$Hg^{2+} + Cl^{-} \rightleftharpoons HgCl^{+}$	$\log K^{\circ}$	7.31	Powell et al. (2005)
$HgCl_2(aq)$	$Hg^{2+} + 2Cl^{-} \rightleftharpoons HgCl_2(aq)$	$\log K^{\circ}$	14	Powell et al. (2005)
$ZrCl^{3+}$	$Zr^{4+} + Cl^{-} \rightleftharpoons ZrCl^{3+}$	$\log K^{\circ}$	1.59	Brown et al. (2005)
$ZrCl_2^{2+}$	$Zr^{4+} + 2Cl^{-} \rightleftharpoons ZrCl_2^{2+}$	$\log K^{\circ}$	2.17	Brown et al. (2005)
PuO_2Cl^{+}	$PuO_2^{2+} + Cl^{-} \rightleftharpoons PuO_2Cl^{+}$	$\log K^{\circ}$	0.23	Guillaumont et al. (2003)
$PuO_2Cl_2(aq)$	$PuO_2^{2+} + 2Cl^{-} \rightleftharpoons PuO_2Cl_2(aq)$	$\log K^{\circ}$	− 1.15	Guillaumont et al. (2003)
$AmCl^{2+}$	$Am^{3+} + Cl^{-} \rightleftharpoons AmCl^{2+}$	$\log K^{\circ}$	0.24	Guillaumont et al. (2003)
$AmCl_2^{+}$	$Am^{3+} + 2Cl^{-} \rightleftharpoons AmCl_2^{+}$	$\log K^{\circ}$	− 0.74	Guillaumont et al. (2003)
$Br_2(l)$		S_f°	152.21	Grenthe et al. (1992)
$TeBr_4(s)$	$Te(s) + 2Br_2(l) \rightleftharpoons TeBr_4(s)$	ΔH_f°	− 190.4	Bard et al. (1985)
$TeI_4(s)$	$Te(s) + 2I_2(s) \rightleftharpoons TeI_4(s)$	ΔG_f°	− 43.4	Borzhim et al. (1979)
		ΔH_f°	− 41.1	Haag et al. (1979)
$Cl^{-}(g)$	$\frac{1}{2}Cl_2(g) + e^{-} \rightleftharpoons Cl^{-}(g)$	ΔH_f°	− 233.3	Hisham and Benson (1992)
$Br^{-}(g)$	$\frac{1}{2}Br_2(l) + e^{-} \rightleftharpoons Br^{-}(g)$	ΔH_f°	− 220.3	Hisham and Benson (1992)
$I^{-}(g)$	$\frac{1}{2}I_2(s) + e^{-} \rightleftharpoons I^{-}(g)$	ΔH_f°	− 197.3	Hisham and Benson (1992)
TeI_6^{2-}	$Te(s) + 3I_2(s) + 2e^{-} \rightleftharpoons TeI_6^{2-}$	E°	0.415	Bigelis et al. (1978)
$BaSO_4(s)$	$BaSO_4(s) \rightleftharpoons Ba^{2+} + SO_4^{2-}$	$\log K^{\circ}$	− 9.95	Brown et al. (2015)
SO_4^{2-}	$S(s) + 2O_2(g) + 2e^{-} \rightleftharpoons SO_4^{2-}$	ΔG_f°	− 744	Grenthe et al. (1992)
HSO_4^{-}	$S(s) + 2O_2(g) + H^{+} + 2e^{-} \rightleftharpoons HSO_4^{-}$	ΔG_f°	− 755.3	Grenthe et al. (1992)
SeO_4^{2-}	$Se(s) + 2O_2(g) + 2e^{-} \rightleftharpoons SeO_4^{2-}$	ΔG_f°	− 439.5	Olin et al. (2005)
$HSeO_4^{-}$	$Se(s) + 2O_2(g) + H^{+} + 2e^{-} \rightleftharpoons HSeO_4^{-}$	ΔG_f°	− 449.5	Olin et al. (2005)

(Continued)

(Continued)

Species	Reaction	Parameter	Value	Reference
HSO_3^-	$S(s) + 1.5O_2(g) + H^+ + 2e^- \rightleftharpoons HSO_3^-$	$\log K°$	7.16	Grenthe et al. (1992)
$HSeO_3^-$	$Se(s) + 1.5O_2(g) + H^+ + 2e^- \rightleftharpoons HSeO_3^-$	$\log K°$	8.36	Olin et al. (2005)
$Ag_2SO_3(s)$	$Ag_2SO_3(s) \rightleftharpoons 2Ag^+ + SO_3^{2-}$	$\log K°$	− 13.82	Sillén and Martell (1964)
$Ag_2SeO_3(s)$	$Ag_2SeO_3(s) \rightleftharpoons 2Ag^+ + SeO_3^{2-}$	$\log K°$	− 15.8	Olin et al. (2005)
$Ag_2TeO_3(s)$	$Ag_2TeO_3(s) \rightleftharpoons 2Ag^+ + TeO_3^{2-}$	$\log K°$	− 17.85	Högfeldt (1982)
Ag^+	$Ag^+ + e^- \rightleftharpoons Ag(s)$	$\Delta G_f°$	77.1	Grenthe et al. (1992)
Pb^{2+}	$Pb^{2+} + 2e^- \rightleftharpoons Pb(s)$	$\Delta G_f°$	− 24.2	Grenthe et al. (1992)
$PbS(s)$	$PbS(s) \rightleftharpoons Pb^{2+} + S^{2-}$	$\log K°$	− 26.6	Ringbom (1953)
$PbSe(s)$	$PbSe(s) \rightleftharpoons Pb^{2+} + Se^{2-}$	$\log K°$	− 42.1	Buketov et al. (1964)
$PbTe(s)$	$PbTe(s) \rightleftharpoons Pb^{2+} + Te^{2-}$	$\log K°$	− 46.3	Buketov et al. (1964)
Hg^{2+}	$Hg^{2+} + 2e^- \rightleftharpoons Hg(l)$	$\Delta G_f°$	164.7	Grenthe et al. (1992)
$HgS(s)$	$HgS(s) \rightleftharpoons Hg^{2+} + S^{2-}$	$\log K°$	− 51.8	Ringbom (1953)
$HgSe(s)$	$HgSe(s) \rightleftharpoons Hg^{2+} + Se^{2-}$	$\log K°$	− 64.5	Buketov et al. (1964)
$HgTe(s)$	$HgTe(s) \rightleftharpoons Hg^{2+} + Te^{2-}$	$\log K°$	− 69.6	Buketov et al. (1964)
Zn^{2+}	$Zn^{2+} + 2e^- \rightleftharpoons Zn(s)$	$\Delta G_f°$	− 147.2	Grenthe et al. (1992)
$ZnS(s)$	$ZnS(s) \rightleftharpoons Zn^{2+} + S^{2-}$	$\log K°$	− 21.6	Ringbom (1953)
$ZnSe(s)$	$ZnSe(s) \rightleftharpoons Zn^{2+} + Se^{2-}$	$\log K°$	− 29.4	Buketov et al. (1964)
$ZnTe(s)$	$ZnTe(s) \rightleftharpoons Zn^{2+} + Te^{2-}$	$\log K°$	− 33.3	Buketov et al. (1964)
Ni^{2+}	$Ni(s) + 2e^- \rightleftharpoons Ni^{2+}$	$\Delta G_f°$	− 44.8	Gamsjäger et al. (2012)
$NiS(s)$	$NiS(s) \rightleftharpoons Ni^{2+} + S^{2-}$	$\log K°$	− 18.5	Ringbom (1953)
$NiSe(s)$	$NiSe(s) \rightleftharpoons Ni^{2+} + Se^{2-}$	$\log K°$	− 32.7	Buketov et al. (1964)
$NiTe(s)$	$NiTe(s) \rightleftharpoons Ni^{2+} + Te^{2-}$	$\log K°$	− 38.1	Buketov et al. (1964)
NO_3^-	$\frac{1}{2}N_2(g) + 1.5O_2(g) + e^- \rightleftharpoons NO_3^-$	$\Delta G_f°$	− 110.8	Grenthe et al. (1992)
$Ag(CN)_2^-$	$Ag^+ + 2CN^- \rightleftharpoons Ag(CN)_2^-$	$\log K°$	19.81	Sillén and Martell (1964)
$CuCl_2^-$	$Cu^+ + 2Cl^- \rightleftharpoons CuCl_2^-$	$\log K°$	5.5	Sillén and Martell (1964)
$Cu(CN)_2^-$	$Cu^+ + 2CN^- \rightleftharpoons Cu(CN)_2^-$	$\log K°$	21.3	Sillén and Martell (1964)
$Hg(CN)_2(aq)$	$Hg^{2+} + 2CN^- \rightleftharpoons Hg(CN)_2(aq)$	$\log K°$	34.71	Sillén and Martell (1971)
$Pd(CN)_4^{2-}$	$Pd^{2+} + 4CN^- \rightleftharpoons Pd(CN)_4^{2-}$	$\log K°$	42.4	Sillén and Martell (1971)
$Pt(CN)_4^{2-}$	$Pt^{2+} + 4CN^- \rightleftharpoons Pt(CN)_4^{2-}$	$\log K°$	41.4	Sillén and Martell (1964)[a]
$PtCl_4^{2-}$	$Pt^{2+} + 4Cl^- \rightleftharpoons PtCl_4^{2-}$	$\log K°$	14.4	Sillén and Martell (1964)
$Zn(CN)_4^{2-}$	$Zn^{2+} + 4CN^- \rightleftharpoons Zn(CN)_4^{2-}$	$\log K°$	19.62	Sillén and Martell (1971)

$Cd(CN)_4^{2-}$	$Cd^{2+} + 4CN^- \rightleftharpoons Cd(CN)_4^{2-}$	log $K°$	17.92	Högfeldt (1982)
CN^-	$\frac{1}{2}N_2(g) + C(s) + e^- \rightleftharpoons CN^-$	$\Delta G_f°$	166.9	Olin et al. (2005)
$H_2TeO_4(aq)$	$HTeO_4^- + H^+ \rightleftharpoons H_2TeO_4(aq)$	$\Delta G_f°$	− 560.2	Calculated from log $K°$
		log $K°$	7.68	Zhdanov (1985)
$HTeO_4^-$	$TeO_4^{2-} + H^+ \rightleftharpoons HTeO_4^-$	$\Delta G_f°$	− 516.4	Calculated from log $K°$
		log $K°$	11.19	Zhdanov (1985)
TeO_4^{2-}	$TeO_4^{2-} + 2H^+ + 2e^- \rightleftharpoons TeO_3^{2-} + H_2O$	$\Delta G_f°$	− 452.5	Calculated from $E°$
		$E°$	0.897	Zhdanov (1985)
$H_2SeO_3(aq)$	$HSeO_3^- + H^+ \rightleftharpoons H_2SeO_3(aq)$	$\Delta G_f°$	− 425.2	Olin et al. (2005)
SeO_3^{2-}	$SeO_3^{2-} + H^+ \rightleftharpoons HSeO_3^-$	$\Delta G_f°$	− 362.4	Olin et al. (2005)
$Ag_2SeO_4(s)$	$Ag_2SeO_4(s) \rightleftharpoons SeO_4^{2-} + 2Ag^+$	$\Delta G_f°$	− 330.2	Olin et al. (2005)
$Ag_2Se(s)$	$Ag_2Se(s) \rightleftharpoons Se^{2-} + 2Ag^+$	$\Delta G_f°$	− 46.9	Olin et al. (2005)
$Ag_2Te(s)$	$Ag_2Te(s) \rightleftharpoons Te^{2-} + 2Ag^+$	$\Delta G_f°$	− 43.1	Wagman et al. (1969)
$PbOH^+$	$Pb^{2+} + H_2O \rightleftharpoons PbOH^+ + H^+$	$\Delta G_f°$	− 218.6	Brown and Ekberg (2016)
$Pb(OH)_2(aq)$	$Pb^{2+} + 2H_2O \rightleftharpoons Pb(OH)_2(aq) + 2H^+$	$\Delta G_f°$	− 401.6	Brown and Ekberg (2016)
$Pb(OH)_3^-$	$Pb^{2+} + 3H_2O \rightleftharpoons Pb(OH)_3^- + 3H^+$	$\Delta G_f°$	− 576.2	Brown and Ekberg (2016)
$Pb_4(OH)_4^{4+}$	$4Pb^{2+} + 4H_2O \rightleftharpoons Pb_4(OH)_4^{4+} + 4H^+$	$\Delta G_f°$	− 927.3	Brown and Ekberg (2016)
$Pb_6(OH)_8^{4+}$	$6Pb^{2+} + 8H_2O \rightleftharpoons Pb_6(OH)_8^{4+} + 8H^+$	$\Delta G_f°$	− 1795.9	Brown and Ekberg (2016)
$Pb(OH)_6^{2-}$	$PbO_2(s) + 2H_2O \rightleftharpoons Pb(OH)_6^{2-} + 2H^+$	$\Delta G_f°$	− 983.5	Brown and Ekberg (2016)
$PbO(s)$	$PbO(s) + 2H^+ \rightleftharpoons Pb^{2+} + H_2O$	$\Delta G_f°$	− 189.3	Brown and Ekberg (2016)
$PbO_2(s)$	$PbO_2(s) + 2H^+ + 2e^- \rightleftharpoons PbO(s) + H_2O$	$\Delta G_f°$	− 218.3	Brown and Ekberg (2016)
$Pb_3O_4(s)$	$Pb_3O_4(s) + 2H^+ + 2e^- \rightleftharpoons 3PbO(s) + H_2O$	$\Delta G_f°$	− 601.6	Robie and Hemingway (1995)
$PbOHCl(s)$	$PbOHCl(s) \rightleftharpoons Pb^{2+} + OH^- + Cl^-$	$\Delta G_f°$	− 390.9	Calculated from log $K°$
		log $K°$	− 13.7	Sillén and Martell (1964)
$Pb(CN)_4^{2-}$	$Pb^{2+} + 4CN^- \rightleftharpoons Pb(CN)_4^{2-}$	$\Delta G_f°$	− 452.5	Calculated from log $K°$
		log $K°$	9.89	Sillén and Martell (1964)[a]
$PbSO_4(s)$	$PbSO_4(s) \rightleftharpoons SO_4^{2-} + Pb^{2+}$	$\Delta G_f°$	− 813.1	Robie and Hemingway (1995) [b]
Au^{3+}	$Au^{3+} + 3e^- \rightleftharpoons Au(s)$	$\Delta G_f°$	440	
$Au(OH)_3(s)$	$Au^{3+} + 3H_2O \rightleftharpoons Au(OH)_3(s) + 3H^+$	$\Delta G_f°$	− 316.9	Yannopoulos (1991)
$HAuO_3^{2-}$	$Au(OH)_3(s) \rightleftharpoons HAuO_3^{2-} + 2H^+$	$\Delta G_f°$	− 142.2	Yannopoulos (1991)[b]
$AuO_2(s)$	$AuO_2(s) + 4H^+ + 4e^- \rightleftharpoons Au(s) + 2H_2O$	$\Delta G_f°$	173.6	
$Au(CN)_2^-$	$Au(CN)_2^- + e^- \rightleftharpoons Au(s) + 2CN^-$	$\Delta G_f°$	285.6	Yannopoulos (1991)
$Ag_2S(s)$	$Ag_2S(s) \rightleftharpoons S^{2-} + 2Ag^+$	log $K°$	− 49.2	Ringbom (1953)

(Continued)

(Continued)

Species	Reaction	Parameter	Value	Reference
$Ag_2O(s)$	$\frac{1}{2}Ag_2O(s) + H^+ \rightleftharpoons \frac{1}{2}H_2O + Ag^+$	log K^o	5.99	Brown and Ekberg (2016)
CH_3COO^-	$2C(s) + 1.5H_2(g) + O2(g) \rightleftharpoons CH_3COO^-$	ΔG_f^o	− 369.4	Bard et al. (1985)
CH_3COOH (aq)	$2C(s) + 2H_2(g) + O2(g) \rightleftharpoons CH_3COOH(aq)$	ΔG_f^o	− 399.6	Bard et al. (1985)
$PbCl_2(s)$	$PbCl_2(s) \rightleftharpoons Pb^{2+} + 2Cl^-$	log K^o	− 4.76	Sillén and Martell (1964)

[a] Corrected to zero ionic strength from data in cited reference.

[b] Fig. 5.5 was constructed to reproduce the pH-potential diagrams given in Marsden and House (2006) and Yannopoulos (1991). The ΔG_f^o value listed in Marsden and House (2006) is 440 kJ mol^{-1}, which is the same value as listed by Bard et al. (1985), but this Gibbs energy doesnot produce the correct pH of conversion between Au^{3+} and $Au(OH)_3(s)$ as indicated by Marsden and House (2006) and Yannopoulos (1991). The hydrolysis of gold(III) is poorly understood (Brown and Ekberg, 2016). Consequently, generic gold species have been used in Fig. 5.5 for the pH of conversion between $Au(OH)_3(s)$ and (1) $Au(OH)_n^{(3-n)+}$ (n = 0–2) and (2) $Au(OH)_{3+n}^{n-}$ (n = 1 or 2) rather than only Au^{3+} and $HAuO_3^{2-}$ and the derived conversion pH is consistent with that indicated by Baes and Mesmer (1976) as well as given by Marsden and House (2006) and Yannopoulos (1991). However, all of these species are not important in the context used herein (see Section 5.4).

References

Baes, C.F., Mesmer, R.E., 1976. The Hydrolysis of Cations. John Wiley and Sons, New York.

Bard, A.J., Parsons, R., Jordan, J. (Eds.), 1985. Standard Potentials in Aqueous Solution. Marcel Dekker Inc, New York.

Bigelis, V.M., Makhkamova, M.Kh, Abrarov, O.A., 1978. Steady-state and equilibrium potentials of tellurium in iodide electrolytes. Electrokhimiya 14, 748–751.

Borzhim, V.S., Levchenko, V.I., Mogil'nitskii, A.A., Prokopovich, L.I., 1979. Determination of standard free energy of tellurium(IV) iodide. Ukr. Khim. Zh. 45, 789–790.

Brown, P.L., Curti, E., Grambow, B., 2005. Chemical thermodynamics of zirconium. In: Mompean, F., Perrone, J., Illemassène, M. (eds), Vol. 8. Elsevier, Amsterdam.

Brown, P.L., Ekberg, C., 2016. Hydrolysis of Metal Ions. Wiley-VCH, Weinheim, 917 p. (2 volumes).

Brown, P.L., Ekberg, C., Ramebäck, H., Hedström, H., Matyskin, A., 2015. Solubility of radium and strontium sulfate across the temperature range of 0 to 300°C. In: Merkel, B.J., Arab, A. (Eds.), Uranium – Past and Future Challenges. Springer, Heidelberg, pp. 553–563.

Buketov, E.A., Ugorets, M.Z., Pashinskin, A.S., 1964. Solubility product and entropy of sulphides, selenides and tellurides. Zh. Neorg. Khim 9, 526–529.

Gamsjäger, H., Gajda, T., Sangster, J., Saxena, S.K., Voigt, W., 2012. Chemical Thermodynamics of Tin. OECD, Paris.

Grenthe, I., Fuger, J., Konings, R.J.M., Lemire, R.J., Muller, A.B., Nguyen-Trung, C., et al., 1992. In: Wanner, H., Forest, I. (Eds.), Chemical Thermodynamics of Uranium. Elsevier, Amsterdam.

Guillaumont, R., Fanghänel, T., Fuger, J., Grenthe, I., Neck, V., Palmer, D.A., et al., 2003. Update on the Chemical Thermodynamics of Uranium, Neptunium, Plutonium, Americium and Technetium. Elsevier, Amsterdam.

Haag, J., Alpen, J.V., Gmelin, E., Rabenau, A., 1979. Electrochemical and specific heat measurements on tellurium-halogen systems. Z. Naturforsch 34a, 969–975.

Hisham, M.W.M., Benson, S.W., 1992. Thermochemistry of inorganic solids. 10. Empirical relations between the enthalpies of formation of solid halides and the corresponding gas-phase halide anions. J. Chem. Eng. Data 37, 194–199.

Högfeldt, E., 1982. Stability Constants of Metal-Ion Complexes. Part A: Inorganic Ligands. Pergamon Press, Oxford.

Lemire, R.J., Berner, U., Palmer, D.A., Tochiyama, O., Musikas, C., Taylor, P., 2013. Chemical Thermodynamics of Iron. Part 1. OECD, Paris.

Lothenbach, B., Ochs, M., Wanner, H., Yui, M., 1999. Thermodynamic Data for the Speciation and Solubility of Pd, Pb, Sn, Sb, Nb and Bi in Aqueous Solution. JNC TN8400 99-011, Japan Nuclear Cycle Development Institute.

Marsden, J.O., House, C.I., 2006. The Chemistry of Gold Extraction, second ed. Society for Mining, Metallurgy and Exploration, Littleton.

McPhail, D.C., 1995. Thermodynamic properties of aqueous tellurium species between 25 and 350°C. Geochim. Cosmochim. Acta 59, 851–866.

Olin, Å., Noläng, B., Osadchi, E.G., Öhman, L.-O., Rosén, E., 2005. Chemical Thermodynamics of Selenium, Vol. 7. Elsevier, Amsterdam.

Panson, A.J., 1963. Polarography of the ditelluride ion. J. Phys. Chem. 67, 2177–2180.

Powell, K.J., Brown, P.L., Byrne, R.H., Gadja, T., Hefter, G., Sjöberg, S., et al., 2005. Chemical speciation of environmentally significant heavy metals with inorganic ligands. Part 1. The Hg^{2+}-Cl^{-}, OH^{-}, CO_3^{2-}, SO_4^{2-} and PO_4^{3-} aqueous systems. Pure Appl. Chem 77, 739–800.

Powell, K.J., Brown, P.L., Byrne, R.H., Gadja, T., Hefter, G., Sjöberg, S., et al., 2007. Chemical speciation of environmentally significant metals with inorganic ligands. Part 2. The Cu^{2+} + OH^-, Cl^-, CO_3^{2-}, SO_4^{2-} and PO_4^{3-} systems. Pure Appl. Chem 79, 895–950.

Powell, K.J., Brown, P.L., Byrne, R.H., Gadja, T., Hefter, G., Leuz, A.-K., et al., 2009. Chemical speciation of environmentally significant metals with inorganic ligands. Part 3. The Pb^{2+} + OH^-, Cl^-, CO_3^{2-}, SO_4^{2-} and PO_4^{3-} systems. Pure Appl. Chem 81, 2425–2476.

Powell, K.J., Brown, P.L., Byrne, R.H., Gadja, T., Hefter, G., Leuz, A.-K., et al., 2011. Chemical speciation of environmentally significant metals with inorganic ligands. Part 4. The Cd^{2+} + OH^-, Cl^-, CO_3^{2-}, SO_4^{2-} and PO_4^{3-} systems. Pure Appl. Chem 83, 1163–1214.

Powell, K.J., Brown, P.L., Byrne, R.H., Gadja, T., Hefter, G., Leuz, A.-K., et al., 2013. Chemical speciation of environmentally significant metals with inorganic ligands. Part 5. The Zn^{2+} + OH^-, Cl^-, CO_3^{2-}, SO_4^{2-} and PO_4^{3-} systems. Pure Appl. Chem 85, 2249–2311.

Ringbom, A., 1953. Solubilities of sulfides. Report to Analytical Section, IUPAC (July).

Robie, R.A., Hemingway, B.S., 1995. Thermodynamic properties of minerals and related substances at 298.15 K and 1 bar (105 pascals) pressure and at higher temperatures. U. S. Geological Survey Bulletin 2131.

Sillén, L.G., Martell, A.E., 1964. Stability Constants of Metal-Ion Complexes. The Chemical Society, London, Special Publication No. 17.

Sillén, L.G., Martell, A.E., 1971. Stability Constants of Metal-Ion Complexes. The Chemical Society, London, Special Publication No. 25.

Wagman, D.D., Evans, W.H., Parker, V.B., Halow, I., Bailey, S.M., Schumm, R.H., 1969. Selected values of chemical thermodynamic properties. Tables for elements 35 through 53 in the standard order of arrangement. U.S. Department of Commerce. National Bureau of Standards. Technical Note 270–274.

Yannopoulos, J.C., 1991. The Extractive Metallurgy of Gold. Van Nostrand Reinhold, New York.

Zhdanov, S.I., 1985. Sulfur, selenium, tellurium, and polonium. In: Bard, A.J., Parsons, R., Jordan, J. (Eds.), Standard Potentials in Aqueous Solution. Marcel Dekker Inc, New York, pp. 93–125.

APPENDIX 2

Polonium Hydroxochloride Complexation

Preliminary absorption spectral studies of polonium complex ion formation in hydrochloric acid were carried out by McCluggage (1949). The equilibrium between the complex ions of polonium responsible for the absorption maxima at 344 and 414 μm were then investigated by Hunt (1954), using an adaptation of the theoretical approach suggested by Bent and French (1941). These studies have been summarized by Moyer (1956).

For equation (A2.1),

$$\mathrm{mA} + \mathrm{nB} \rightleftharpoons \mathrm{A_m B_n} \tag{A2.1}$$

the equilibrium constant is

$$K = \frac{[A_m B_n]}{[A]^m [B]^n} \tag{A2.2}$$

and hence

$$\log\,[A_m B_n] = m \log\,[A] + n \log\,[B] + \log \mathrm{K} \tag{A2.3}$$

If $\mathrm{A}_{[A_m B_n]}$ represents the absorbance due to $[A_m B_n]$, then from Beer's law,

$$\mathrm{A}_{[A_m B_n]} = \varepsilon\, l\, [A_m B_n] \tag{A2.4}$$

where ε is the molar absorbtivity and l is the cell thickness. Taking logarithms, Eq. (A2.4) becomes

$$\log\,[A_m B_n] = \log A_{[A_m B_n]} - \log \varepsilon\, l \tag{A2.5}$$

Eq. (A2.3) can, therefore, be rewritten as

$$\log \mathrm{A}_{[A_m B_n]} = m \log\,[A] + n \log\,[B] + \mathrm{C} \tag{A2.6}$$

where C is a constant equal to $\log K + \log \varepsilon\, l$.

If the absorbance of the complex, $\log A_{[A_m B_n]}$, is plotted against the logarithm of the concentration (at equilibrium) of one of the other reactants, with the concentration of the other reactant held constant, a straight line should result, the slope of which is equal to the stoichiometric coefficient, *m* or *n*, of the reactant in the complex.

References

Bent, H.E., French, C.L., 1941. The structure of ferric thiocyanate and its dissociation in aqueous solution. J. Am. Chem. Soc. 63, 568–572.

Hunt, D.J., 1954. Polonium Complexes in Chloride Solutions by Absorbancy Studies. Mound Laboratory Report, MLM-979.

McCluggage, W.C., 1949. Quarterly Progress Report. Mound Laboratory Report, MLM-405-2, p. 71.

Moyer, H.V., 1956. Chemical properties of polonium. In: Moyer, H.V., Gnagey, L.B., Rogers, A.J. (Eds.), Polonium. U.S. Atomic Energy Commission, pp. 33–96. TID-5221.

Index

Note: Page numbers followed by "*f*" and "*t*" refer to figures and tables, respectively.

A

Acetylacetone, 24–25, 32–33
Aciding and silver recovery, 147–150
  Acid leaching experiments
  with different leachants, 157–159, 158*t*
  sulfuric acid leaching, 155–156, 157*t*
Activated carbon, 163–164
Adsorption, 35–37
  enthalpy and entropy, 21
Ag_2PoO_3(s), 96–98
Anglesite ($PbSO_4$), 128, 136–137, 157–158
Anion exchange, 22, 35
Anode slimes processing, 130–162, 134*f*
  aciding and silver recovery, 147–150
  copper anode (raw) slimes, 134–136
  cyanidation, 135*f*, 139–142
  decopperization, 132, 136–139
  at Olympic Dam, South Australia, 134*f*
  phases in solids, 136*t*
  polonium behavior during, 130–162
  solids and liquors, 135*t*
  zinc precipitation, 135*f*, 143–147
Aqueous speciation, 128–129
Autodeposition, 168–170

B

Baird grating spectrograph, 9
Basic anion exchange adsorption, 35
Beryllium polonide (BePo), 25
BHP's Olympic Dam Operations at Roxby Downs, 4–5
Bismuth, 134–136, 139–140
Bismuth polonide (BiPo), 26
Bone marrow syndrome, 2
Bornite, 130

C

Cadmium polonide (CdPo), 26
Calcium hypochlorite, 152–153
Carbon polonide (CPo), 29
Cataphoresis, 36
Cementation, 144
Central nervous system syndrome, 2
Cerrusite ($PbCO_3$), 128
Chalcocite, 130
Chalcopyrite, 130
Chemical properties
  elemental polonium, 10
  isotopes, 7–8
  oxidation states, 10–15
  polonides, polonites, and polonium compounds with other chalcogens, 27–29
  polonium halides, 18–25
  polonium ion exchange, 33–36
  polonium nitrates, 29–31
  polonium oxides, hydroxides and hydrides, 16–18
  solvent extraction of polonium, 31–33
  tracer experiments, 29, 37
Chemical thermodynamics
  ionic strength corrections, 48–52
  polonium species, thermochemical properties for, 52–116, 110*t*
    Ag_2PoO_3(s), 96–98
    H_2Po(aq) and HPo^-, 65–68
    $HPoO_3^-$, 60–63
    H_2PoO_3(aq), 58
    metal polonides, 101
    organic complexes, 107–116
    PbPo(s), HgPo(s), ZnPo(s), NiPo(s) and Ag_2Po(s), 98–101
    Po^{2-}, 53–55
    Po^{2+}, 56
    $PoBr_2$(s), 83
    $PoBr_4$(s), 83–84
    $PoCl_4^{2-}$, 71–72
    $PoCl_6^{2-}$, 76–78

Chemical thermodynamics (*Continued*)
 $PoCl^+$, $PoCl_2(aq)$ and $PoCl_3^-$, 73–75
 $PoCl_2(s)$, 68–69
 $PoCl_4(s)$, 69–70
 $Po(CN)_6^{2-}$, 104–105
 PoI_5^- and PoI_6^{2-}, 85
 $PoI_2(s)$ and $PoI_4(s)$, 84
 PoO_3^{2-}, 58–59
 PoO^{2+} and $PoOOH^+$, 63–65
 $PoOCl_4^{2-}$ and $PoO(OH)Cl_2^-$, 82–83
 $PoO(CN)_2(s)$, 105–107
 $PoOHCl_4^-$, 78–81
 $PoONO_3^+$, $PoO(NO_3)_2(aq)$, and $PoO NO_3)_3^-$, 102–104
 $(PoO)_2(NO3)_3OH(s)$, 101–102
 $(PoO)_2OSO_4(s)$, 91–92
 $PoO_2(s)$, 56–58
 $PoO_3(s)$, 65
 $PoOSeO_4(aq)$ and $PoO(SeO_4)_2^{2-}$, 95–96
 $(PoO)_2SeO_4(s)$, 93–95
 $PoOSO_4(aq)$ and $PoO(SO_4)_2^{2-}$, 87–88
 $Po(s)$, $Po(g)$ and $Po_2(g)$, 52–53
 $Po(SO_4)_2$. $H_2O(s)$ and $PoO(SO_4)_3^{4-}$, 88–90
 $PoSO_4(aq)$, 92–93
 $PoSO_4(s)$, 86–87
 $PoS(s)$, 85–86
 principles, 43–52
Chloroplatinates, 21
Chloroplumbates, 21
Chlorotellurites, 21
Cocrystallization, 15
Colloidal properties, 36
Copper anode slimes, 134–136
 hydrometallurgical treatment, 133
Coprecipitation, 161
 experiments, 21
Critical deposition potential, 13
Cubic fluorite (CaF_2), 15
Cyanidation, 133, 135*f*, 139–142
Cyanided anode slimes, oxidation of, 153–154, 153*t*, 154*t*
Cyclotrons, 1–2

D

Decopperization, 132–133, 136–139
Decopperized slimes, 140
Diethylammonium diethyldithiocarbamate (DDTC), 33
Diethyldithiocarbamate (DTC), 33
Disulfate phases, 27
Dithizone, 31–32
Dysprosium polonide (DyPo), 29

E

Electrochemical deposition experiments, 11
Electrode potentials, 13, 14*t*
Elemental polonium, 8–10
 chemical properties, 10
 physical properties, 8–10
Enthalpy, 21, 46, 101
Entropy, 47–48, 101
Equilibrium constant, 17, 23–24, 122
Extraction from bismuth, 32

F

Face-centered cubic structure, 18–19, 25
Ferric iron, 168
Fischesserite, 145–147
Free cyanide, 152
"Free energy of reaction", 44

G

Gadolinium polonide (GdPo), 29
Galena (PbS), 128
Gastrointestinal syndrome, 2
Germanium polonide (GePo), 29
Gibbs energy, 45–46, 61–65, 101
Gibbs energy of formation, 43–44

H

Hafnium polonide (HfPo), 29
Health issue, 2
Hematopoietic latency, 2
Hexachloropolonites, 18–19
$H_2Po(aq)$ and HPo^-, 65–68
$HPoO_3^-$, 60–63
$H_2PoO_3(aq)$, 58
Hydrated dioxide, 16, 27

Hydrated disulfate, 27
Hydrochloric acid leach, 173–174
Hydrochloric acid solution, 20–21
Hydrogen polonide (H_2Po), 18
Hydrogen telluride, 18
Hydroxylamine, 11, 20–21
Hypophosphorous acid, 11

I

Indium polonide (InPo), 29
Iodine free hydroiodic acid, 16
Ionic radii, 9–10
Isotopes, 1, 7–8

L

Lanthanum polonide (LaPo), 29
Lead, 179
  aqueous speciation, 128–129
  dissolution, 157
  ore, 128
Lead-210, 166
Lead-bearing rooseveltite, 136
Lead – chlorine – water system, 157, 159*f*, 167, 167*f*, 174–176, 175*f*
Lead – cyanide – water system, 141, 142*f*, 145*f*
Lead-hydroxide complexes, 141
Lead/polonium sulfate, preparation, 160–162
Lead sulfide, 174
Lead – sulfur – water system, 148, 149*f*
Lead – water system, 130*f*, 164, 164*f*, 171–172, 172*f*
Lethal dose (LD_{50}), 2
Linear free energy, 61–62, 75*f*
Liquid-phase adsorbents, 163–164

M

Magnitude, 2
Mercury polonide (HgPo), 26
Mesityl oxide, 32
Metal chalcogenide solubilities, 99*t*
Metal chloride complexes, 74*t*
Metallic tellurium, 36
Metal polonides, 29, 98–99, 101
Methylisobutylketone (MIBK), 12, 24–25, 32–33
Minerals, solid solutions, 161
Molar absorptivity, 78–79
Monazite, 170–173
Monopolonide phases, 25–26
Monoxide, 16
Montmorillonite (clay), 37

N

Naumannite, 139–140, 145–147
Neodymium polonide (NdPo), 29
Neutral polonium atoms, 8
Nickel polonide (NiPo), 26
Nitrites, 11
Nitrogen analysis, 20
Nonelectrochemical reactions, 45
Normal calomel electrode, 13, 15
Nuclear reactors, 1–2

O

Occlusion, 161
Organic complexes of polonium, 107–116
Oxalic acid, 11, 20–21
Oxidation experiments, pH-potential diagram
  using calcium hypochlorite, 152–153
  using sodium hypochlorite, 154–155
Oxidation states, 10–15

P

PbPo(s), HgPo(s), ZnPo(s), NiPo(s) and Ag_2Po(s), 98–101
pH – potential diagram
  acid leaching experiments, 155–159
  construction, 122–126
  derivation, 127–128
  domains, 123*t*
  lead – chlorine – water system, 157, 159*f*, 167, 167*f*, 174–176, 175*f*
  lead – cyanide – water system, 141, 142*f*, 145*f*
  lead/polonium sulfate, preparation, 160–162
  lead – sulfur – water system, 148, 149*f*

pH – potential diagram (*Continued*)
  lead – water system, 164, 164*f*, 171–172, 172*f*
  oxidation experiments, 152–155
  polonium, aqueous speciation for, 162
  polonium – chlorine – sulfur – water system, 176, 176*f*
  polonium – chlorine – water system, 157, 158*f*, 165, 165*f*, 168, 169*f*, 173, 174*f*
  polonium – cyanide – water system, 141, 141*f*, 144–145, 144*f*
  polonium – silver – chloride – water system, 169–170, 170*f*
  polonium – sulfur – water system, 137*f*, 147, 148*f*
  polonium – water system, 122, 123*f*, 128, 163*f*, 172, 173*f*
  selenium – water system, 145–150, 146*f*, 149*f*, 151*f*, 158–159, 159*f*
  silver – selenium – water system, 138–139, 138*f*
  silver – tellurium – water system, 138–139, 139*f*
  tellurium – water system, 129*f*, 142, 143*f*, 145–150, 146*f*, 150*f*, 151*f*, 158–159, 160*f*
Physical properties
  adsorption and desorption experiments, 37–38
  colloidal properties, 36–37
  elemental polonium, 8–10
  isotopes, 7–8
  polonides, polonites, and the polonium compounds with other chalcogens, 25–26
  polonium halides, 18–19
  polonium ion exchange, 33–36
  polonium oxides, hydroxides and hydrides, 15
  solvent extraction of polonium, 31–33
Platinum group metals (PGMs), 130
Platinum polonide (PtPo), 26
Po^{2-}, 53–55
Po^{2+}, 56
$PoBr_2(s)$, 83
$PoBr_4(s)$, 83–84
$PoCl_4{}^{2-}$, 71–72
$PoCl_6{}^{2-}$, 76–78
$PoCl^+$, $PoCl_2(aq)$ and $PoCl_3{}^-$, 73–75
$PoCl_2(s)$, 68–69
$PoCl_4(s)$, 69–70
$Po(CN)_6{}^{2-}$, 104–105
$PoI_5{}^-$ and $PoI_6{}^{2-}$, 85
$PoI_2(s)$ and $PoI_4(s)$, 84
Poisoning, 2–3
Polonide phases, 25–26
Polonides, polonites, and polonium compounds with other chalcogens, 25–29
  chemical properties, 27–29
  physical properties, 25–26
Polonium-208, 7
Polonium-210, 1, 122, 131, 153, 163, 166
  extraction, 33
  in tobacco, 2
  toxicity, 2
Polonium chloride complexes, 79*t*, 80*t*
Polonium – chlorine – sulfur – water system, 176, 176*f*
Polonium – chlorine – water system, 157, 158*f*, 165, 165*f*, 168, 169*f*, 173, 174*f*
Polonium – cyanide – water system, 141, 141*f*, 144–145, 144*f*
Polonium dibromide, 19, 22
Polonium dioxide (PoO_2), 15–16
Polonium halides, 18–25
  chemical properties, 19–25
  physical properties, 18–19
Polonium hydroxochloride complexation, 191–192
Polonium(IV) in perchloric acid solutions, 32
Polonium ion exchange, 33–36
Polonium(IV) sulfate, 27
Polonium nitrates, 12
  chemical properties, 29–31
Polonium oxides, hydroxides and hydrides, 15–18
  chemical properties, 16–18
  physical properties, 15
Polonium selenate, 28–29
Polonium – silver – chloride – water system, 169–170, 170*f*

Polonium sulfide, 7–8
Polonium – sulfur – water system, 137*f*, 147, 148*f*
Polonium tetra- and dichlorides, 21
Polonium tetrabromide, 22
Polonium tetrachloride, 18, 20–21, 27–29
Polonium tetrafluoride, 24
Polonium tetraiodide, 19, 22–23
Polonium – water system, 123*f*, 128, 163*f*, 172, 173*f*
PoO_3^{2-}, 58–59
PoO^{2+} and $PoOOH^+$, 63–65
$PoOCl_4^{2-}$ and $PoO(OH)Cl_2^-$, 82–83
$PoO(CN)_2(s)$, 105–107
$PoOHCl_4^-$, 78–81
$PoONO_3^+$, $PoO(NO_3)_2(aq)$, and PoO $NO_3)_3^-$, 102–104
$(PoO)_2(NO3)_3OH(s)$, 101–102
$(PoO)_2OSO_4(s)$, 91–92
$PoO_2(s)$, 56–58
$PoO_3(s)$, 65
$PoOSeO_4(aq)$ and $PoO(SeO_4)_2^{2-}$, 95–96
$(PoO)_2SeO_4(s)$, 93–95
$PoOSO_4(aq)$ and $PoO(SO_4)_2^{2-}$, 87–88
Po(s), Po(g) and $Po_2(g)$, 52–53
$Po(SO_4)_2$. $H_2O(s)$ and $PoO(SO_4)_3^{4-}$, 88–90
$PoSO_4(aq)$, 92–93
$PoSO_4(s)$, 86–87
PoS(s), 85–86
Process liquors, 5
Prodrome, 2
Properties of polonium, 9*t*
  chemical properties. *See* Chemical properties
  physical properties. *See* Physical properties

R

Radioactivity, 7–8
Radiolead nitrate, 37–38
Radio-tellurium, 1
Radium-E, 1
Radium-F, 1
Rare earth elements (REEs), 170
mineral processing, 170–176
Raw slimes, mineralogical analyses of, 136

S

Seawater, polonium in, 165–168
Selenium, 8, 132, 179
  aqueous speciation, 128–129
Selenium – water system, 142, 143*f*, 145–150, 146*f*, 149*f*, 151*f*, 158–159, 159*f*
Silver and gold electrorefining
  polonium behavior during, 163–165
Silver polonide, 26
Silver – selenium – water system, 138–139, 138*f*
Silver – tellurium – water system, 138–139, 139*f*
Sodium cyanide, 139–140
Sodium dithionite, 11
Sodium hypochlorite, 154–155
Sodium polonide, 25, 29
Solid phase speciation, 138–139
Solid solutions, 161
Solvent extraction, 31–33
  techniques, 19
Stability constant, 32, 77, 87–88, 96, 107
Static charges elimination, 3
Sulfate complexation, 32
Sulfide minerals, 130
Sulfuric acid, 27
Sulfuric acid leaching, 155–156, 157*t*
Surface adsorption, 161

T

Tellurium, 8, 179
  aqueous speciation, 128–129
  removal, 132
Tellurium (IV) sulfate, 28
Tellurium – water system, 129*f*, 142, 143*f*, 145–150, 146*f*, 150*f*, 151*f*, 158–159, 160*f*
Thenoyltrifluoroacetone (TTA), 109
Thermochemical database, 4–5
Thermoelectric generator, 3
Tin polonide (SnPo), 29
Titanium polonide (TiPo), 29
Tracer solution chemistry, 16–17
Tributyl phosphate (TBP), 25
Trioxide, 16
Trivalent titanium, 11

U

Uranium-238, 122
Uranium-rich ores, 3–5

V

Volatile polonium fluoride compound, 24
Volatilized polonium, 179

X

X-ray diffraction (XRD) studies, 8

Z

Zero ionic strength, 64, 74*t*, 87–88, 103, 104*t*
Zinc precipitation, 133, 135*f*, 143–147
Zirconium polonide (ZrPo), 29

Kolibri 360 GmbH
Ferdinand-Jühlke-Straße 7
99095 Erfurt, Deutschland
produktsicherheit@kolibri360.de